MAPPING THE SKY

THE ESSENTIAL GUIDE TO ASTRONOMY

BY LEÏLA HADDAD & ALAIN CIROU

ILLUSTRATIONS BY MANCHU

SEUIL CHRONICLE

Contents

MAPPING THE SKY

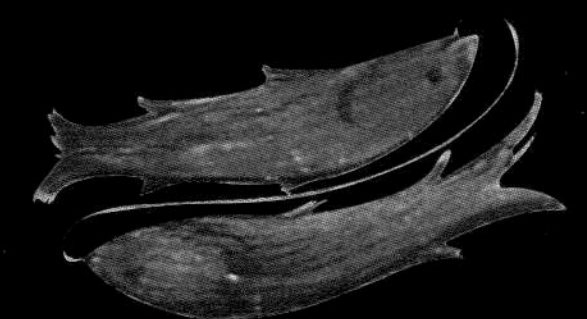

Part Two

[Paris, BNF. © BNF]

To understand astronomy, should you begin by observing the sky? Or should you initiate yourself with theory before lifting your eyes to the heavens? Neither one nor the other . . . and both at the same time!

Basic astronomy, the science that describes, explains and predicts the movements of celestial bodies, inextricably combines observation and theory. At the beginning of the seventeenth century, German astronomer Johannes Kepler only succeeded in unveiling the secret behind planetary movement in our solar system thanks to the piercing gaze of the sixteenth-century Dane Tycho Brahe. An exceptional observer, Brahe spent practically every night of his life with his eyes riveted to the celestial vault. He scrupulously and minutely studied the path of Mars and his measurements led Kepler to realize that the planets follow elliptical trajectories around the sun. More than two centuries later, in France, Urbain Le Verrier discovered an eighth lodger in our solar system without even needing to set foot outside: he required only a pencil, a sheet of paper, Isaac Newton's theory of universal attraction, and his own undeniable mathematical genius. The planet Neptune was observed for the first time in 1846, in the exact spot foreseen in Le Verrier's calculations.

Observation and theory mutually enrich each other: sight fuels reflection, which gives meaning to what we look at. This perpetual give-and-take, so characteristic of astronomy, changes our perception of the world: what we observe may prove to be the direct opposite of what we know. We see the sun rise and set, and everything in its apparent movement leads us to believe that it is in orbit around the earth. But we know it's the earth that moves . . .

This book refuses to choose between seeing and knowing. The first section shows how knowledge was constructed and how, little by little, we came to understand the veritable march of the stars, how we forged the tools that allowed us to draw up celestial charts, to know the present, past and future behavior of celestial bodies, and to domesticate time.

The second part of this book is a guide for observation. It unlocks, one by one, the mysteries of the sky, from the art of getting oriented amidst the constellations to crossing the frontiers of the distant universe with the aid of binoculars and telescopes. Pleasure and curiosity are the driving forces behind this plunge into darkness. Between knowing and seeing, we are all free to navigate as we please, following our desires, needs and dreams.

LEÏLA HADDAD AND ALAIN CIROU

ASTRONOMY EMERGED four thousand years ago from ancient Mesopotamia, between the Tigris and Euphrates rivers. It became a science in Greece in the fifth century B.C. Its intention was to study the movements of celestial bodies. Its goal was to clear the sky for cultivation, establish celestial charts, understand time, follow the stars in their age-old travels and provide their past, present and future positions as precisely as possible. The beautiful edifice of basic astronomy rests upon a few simple laws—the logical, geometric and physical pillars of the cosmos. They are accessible to all who wish to know the sky, through the rich history of their genesis, where imagination, observation and simple common sense have starring roles.

Leïla Haddad

UNLOCKING ASTRONOMY

LEÏLA HADDAD

Naples, Biblioteca nazionale Vittorio Emanuele III

Part One

1
The Celestial Sphere

1. THE WORK OF THE GODS 2. THE WORLDLY SPHERE 3. THE SPHERE OF FIXED STARS

"Henceforth I spread confident wings to space. I fear no barrier of crystal or of glass; I cleave the heavens and soar to the infinite. And while I rise from my own globe to others and penetrate ever further through the eternal field, that which others saw from afar, I leave far behind me." At the end of the sixteenth century, the excommunicated monk Giordano Bruno, poet, philosopher, physicist and occasional magician, became the first person to dare proclaim what seems obvious to us today: that the universe is infinite. Throughout the centuries that followed his suicidal liftoff—accused of heresy, he was burned alive in Rome in February 1600—the closed universe inherited from antiquity, as smooth, dense and firm as a billiard ball, literally exploded. It became "space," an elsewhere that can only be described in abstract mathematical equations, a sort of dark, bottomless well, without light and without walls, so breathtakingly deep and indeterminate that to the naked eye it looks like nothing at all.

The universe, however, did have a shape. Until the fourth century B.C., our image of it was purely mythological: the gods had fashioned it according to their whim, according it any form—house, tree, den, dome—that pleased them. According to Greek thinkers and mathematicians, the universe became an enormous sphere. The earth was the unmoving navel of this gigantic celestial marble, outside of which nothing existed. This roundish conception of the universe was not gratuitous, but was partially deduced by observing the trajectories of celestial bodies. In becoming spherical, moreover, the sky acquired the properties of what was now considered the most perfect geometrical figure, which allowed a veritable science of the stars to take off.

This vision of the universe, almost perfect in the degree to which it was simple and practical, persisted until the sixteenth century. Then astronomy had its revolution, the earth began to turn, and the universe revealed what seemed to be its true face. Astronomers did not bury the sphere as such: to answer the needs of astrometry—the branch of astronomy devoted to studying the movements of celestial bodies—they continued to treat the stars like butterflies pinned to the inside wall of a giant ball as it rotated around a fixed earth. The sphere became a geometrical model, a clever and brilliant trick that allowed astronomers to represent much of how the universe worked without needing to lose themselves in its depths.

According to Greek mythology, Zeus had nine daughters with Mnemosyne, the personification of memory. These were the famous Muses of antiquity, protectors of the arts, letters and sciences. Urania—her name came from Ouranos, the sky—was the kindly patroness of astronomy.
[Engraving, seventeenth century (detail). Paris, BNF. © BNF]

1.

THE WORK OF THE GODS

Whatever its culture or religion, every civilization has resorted to myth at one point or another in its history. In the absence of science, myth alone can explain and justify the existence of everything in the universe—humankind, the stars, the moon, the sun, the earth, the sky, death, love, thunder, trees, wheat, wind. The responsibility of creating and organizing the universe is conferred upon entities with colossal abilities and powers: the gods, omnipotent creatures who naturally steer all their attention toward earth. How could it have been otherwise? Without rival in all creation, the earth stretches out solid, massive and nourishing beneath the feet of humankind. The earth is the *raison d'être* of the cosmos, which closes itself around her like an oyster on a pearl.

The Chinese demiurge Pan Gu, the creator of the world, died after separating the sky from the earth. The different parts of his body were then transformed into mountains, seas, rivers, trees, stones, metals . . .
[© Jean-Louis Charmet]

THE CREATION OF THE WORLD

According to Chinese cosmogony, the mythic tale of the creation of the world, an egg lay at the origin of everything. Having emerged from the great primordial nothing, it contained the seeds of the world to come, in the form of the entity Pan Gu. This entity began to grow and grow until, eighteen thousand years after its appearance, the egg finally cracked. Its brightest and lightest fragments rose up to form the celestial vault, while its heaviest and darkest elements sank to form the earth. Pan Gu continued to grow, forcing the sky to become more and more distant from the earth. At the end of a second eighteen-thousand-year cycle, both had solidified into their current positions, and Pan Gu could finally die. The rest of creation was forged from the different parts of his body: his right eye became the moon, his left produced the sun, his voice became thunder and his hair, the stars.

In Egyptian mythology, Atum is at the origin of the world. Sitting on a hillock in the middle of Nun, the primordial waters, he engaged in solitary pleasure and from his semen—or his spittle—spurted forth Shu and his sister Tefnut. The two begat Nut and Geb, who entered the world so tightly wrapped together that Nut, pregnant with four children, could not give birth for lack of space. Shu then slipped between them and separated them. Nut became the sky, beautifully arched over her brother's stretched out body, the earth.

In more brutal fashion, the Vikings believed the world was born from the dismemberment of a giant named Ymir. His body became the earth, his skull became the sky and the ocean spewed forth from his blood.

THE ORGANIZATION OF THE SKY

In the everyday affairs of the sky the Mesopotamians took divine interference to the limit. These formidable observers spent more than two thousand years on the lookout for anything that might happen in the firmament, and were perfectly aware of any peculiarity in the movements of the stars. And yet, they never sought to explain them. To the Mesopotamians, the world had been conceived, down to its tiniest details, by the god Marduk, grandson of the primordial goddess Tiamat, whom he defeated in single combat. He sliced her body

In North America, India, Tibet and China, many believed that the earth was carried by an enormous tortoise. The dorsal and ventral shell of the slow creature symbolized the sky and the subterranean world, between which rested the tender flesh of the earth. To some the serpent represents the dark forces of chaos.

[Mount Meru, the earth and underworlds carried by a tortoise, India, nineteenth century. © Jean-Louis Charmet]

into two halves, the first for the earth and the second for the celestial vault, which he organized from A to Z. He carefully surveyed and measured it, to determine the placement of a gigantic palace where the gods would live. He then arranged the stars, replicas of the gods, into constellations and traced across the sky the route the sun would take a year to travel.

He ordered Jupiter to oversee the proper working of the sky, opened the eastern and western doors through which the sun appears and disappears, placed the zenith point in Tiamat's womb and, finally, created Nannar, the moon, to whom he entrusted time. He described her task in minute detail, fixing the days and

According to the Mesopotamians, the world comprises two shells, the earth and the sky, hermetically sealed and submerged in an infinite ocean. The third world, that of the dead, is housed at the center of the earth. This universe was entirely created and organized by the god Marduk, who even thought of opening two doors to the east and the west to permit the sun to enter and leave the sky.

[The sky and earth according to the Babylonians, illustration by P. Jensen, late nineteenth century. Paris, Bibliothèque des Arts décoratifs. © Jean-Louis Charmet]

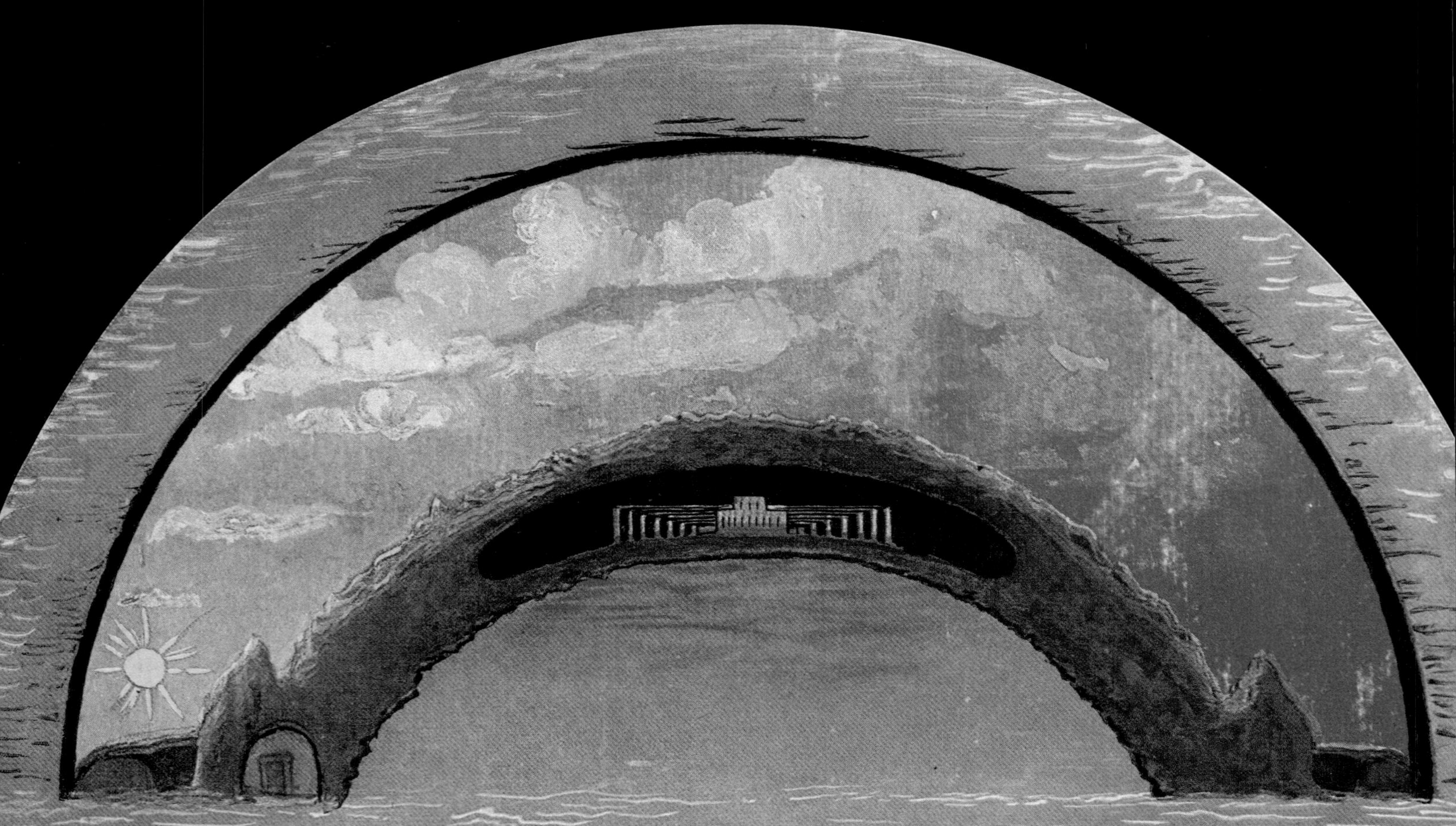

the various guises in which she should reveal herself to the world. In short, the universe is how it is because this was the will of Marduk. And it was utterly useless, and even sacrilegious, to question it.

THE MYTHICAL ARCHITECTURE OF THE COSMOS

Whatever its origin—egg, divine spittle, or sublime butchery—the mythic universe closely resembles a clam. The earth, a little lump of palpitating, fragile flesh, sometimes imagined as an infinite, circular tray surrounded by water, usually finds itself wedged between two gigantic, solid shells. The first shell, the sky, is the upper world, where the majority of the gods have chosen to reside. The second, the lower world, belongs to the obscure legions of Death. This vision of a universe constructed like a hermetically sealed box or oyster is common to a great number of cultures. It can be found in Asia, North America and Europe, with occasional variations: for the Mongols, the upper and lower shells contained several levels across which shamans could circulate. The Norse people gave the world an axis, the giant tree Yggdrasil, whose roots dove deep into the earth and whose highest leaves tickled the heavens. As for the Egyptians, they took the shell analogy so far as to submerge their cosmos into the chaotic ocean of Nun.

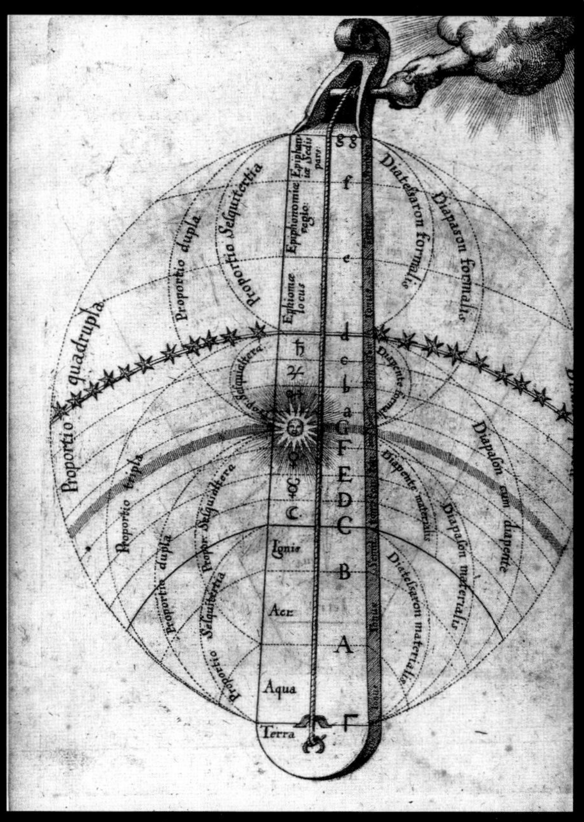

The cosmos as a musical instrument with one string, tuned by the hand of God? Pythagoras and his disciples discovered that the sound emitted by a lyre depended on the length of its strings. By analogy, they were convinced that there was a secret and mathematical order to the cosmos

[Robert Fludd, *Utriusque cosmi, majoris scilicet et minoris, metaphysica, physica atque technica historia* Paris, BNF © BNF]

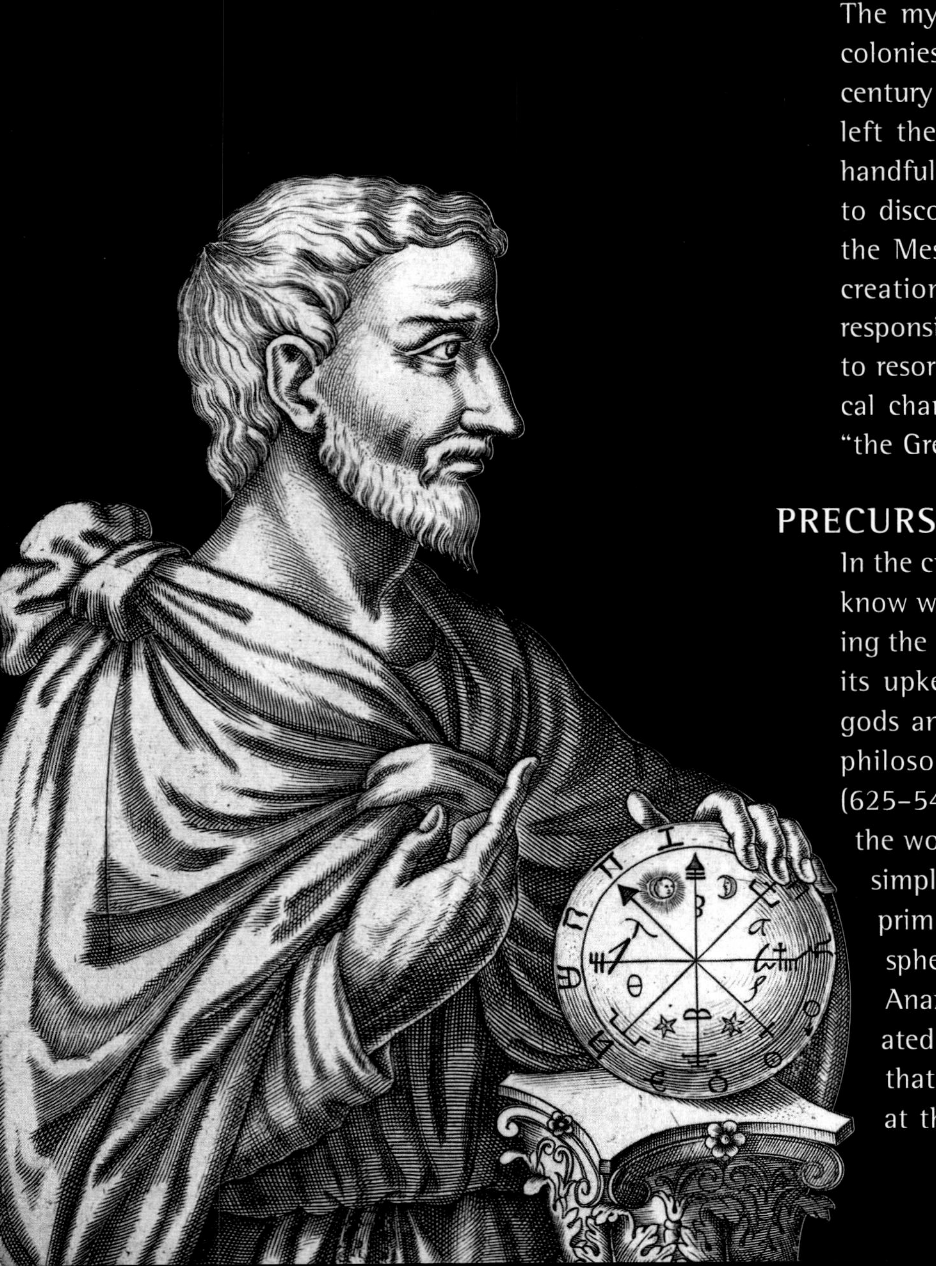

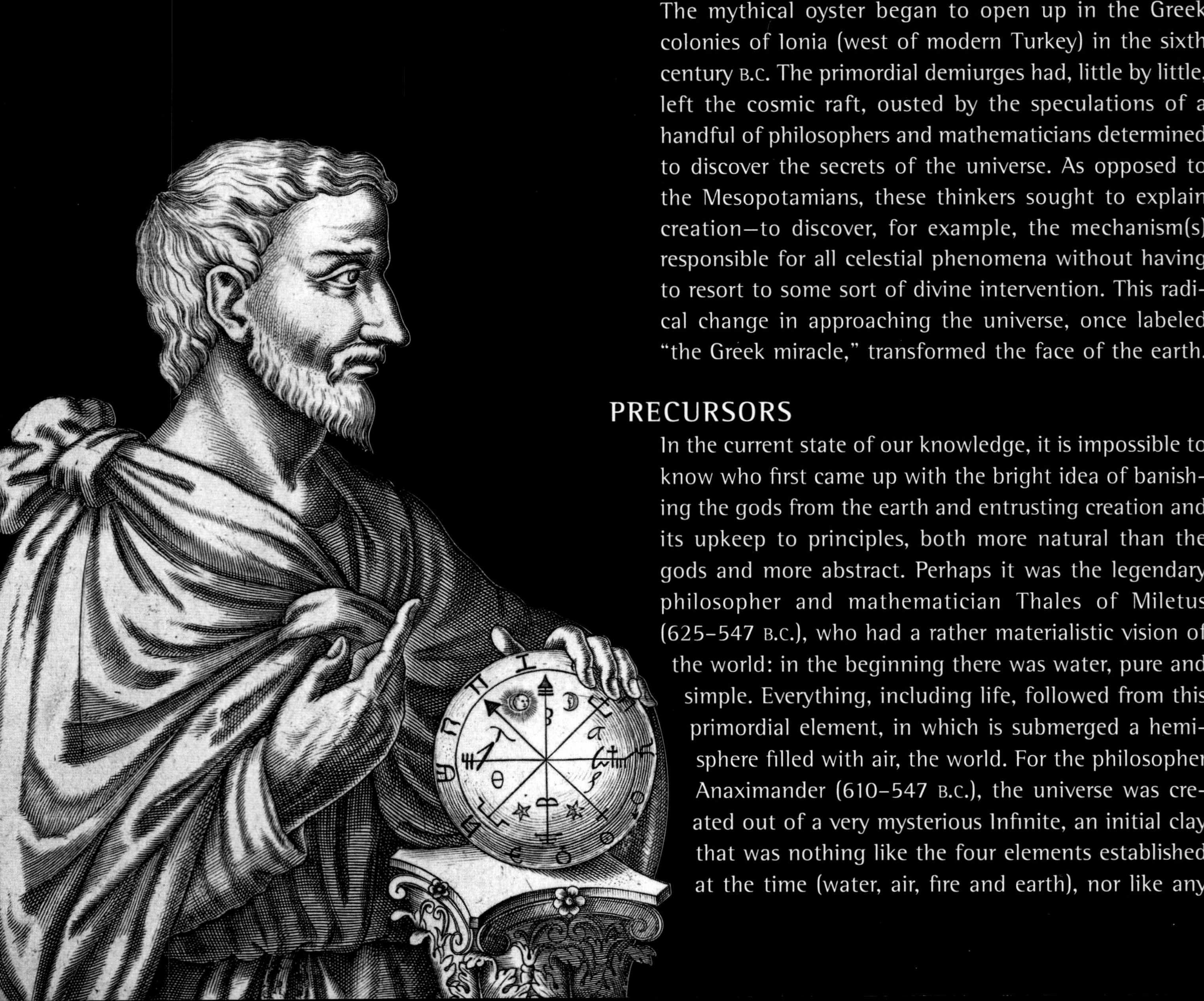

The mythical oyster began to open up in the Greek colonies of Ionia (west of modern Turkey) in the sixth century B.C. The primordial demiurges had, little by little, left the cosmic raft, ousted by the speculations of a handful of philosophers and mathematicians determined to discover the secrets of the universe. As opposed to the Mesopotamians, these thinkers sought to explain creation—to discover, for example, the mechanism(s) responsible for all celestial phenomena without having to resort to some sort of divine intervention. This radical change in approaching the universe, once labeled "the Greek miracle," transformed the face of the earth.

PRECURSORS

In the current state of our knowledge, it is impossible to know who first came up with the bright idea of banishing the gods from the earth and entrusting creation and its upkeep to principles, both more natural than the gods and more abstract. Perhaps it was the legendary philosopher and mathematician Thales of Miletus (625–547 B.C.), who had a rather materialistic vision of the world: in the beginning there was water, pure and simple. Everything, including life, followed from this primordial element, in which is submerged a hemisphere filled with air, the world. For the philosopher Anaximander (610–547 B.C.), the universe was created out of a very mysterious Infinite, an initial clay that was nothing like the four elements established at the time (water, air, fire and earth), nor like any other substance known on earth. Anaximander was the first to take our planet out of its shell by asserting it was much like a cylinder—far more wide than high—that held itself in place, perfectly immobile, at the center of the universe. As for the stars, he compared them to sorts of wheels filled with fire, through whose centers streams light. When they get clogged up, we see an eclipse.

E PYTHAGOREANS

The world became round thanks to the sectlike school of philosophy founded around 530 B.C. in Crotone, in southern Italy, by the mythic mathematician Pythagoras. His disciples had a special fondness for the sphere, which they considered the most perfect geometric form conceivable. Thrilled by the discovery that the earth was round, perhaps after observing the shape of its shadow during lunar eclipses, they declared that the five planets known at the time—Mercury, Venus, Mars, Jupiter and Saturn—were spheres as well. Why not? Long and attentive observation of the night sky leads one to the realization that they all seem to rotate around the earth, following paths that the Pythagoreans, ascetics keen on order, perfection and simplicity, argued were circular. The Pythagoreans were convinced that everything in the world is associated with a number. To discover the laws of the universe, you need only look for relationships between numbers. This belief was supported by one of their many finds, such as that the sound emitted by the string of a lyre depended upon its length. Clearly, music has nothing to do with the instrument; it is merely a combination of numbers. Gliding around the earth, the celestial marbles, too, had to emit a sound, a function of the length of their trajectory and speed. According to legend, only Pythagoras

Pythagoras, between the sixth and fifth centuries B.C., supposedly declared that the earth and universe could only have the form of a sphere, because it was the most perfect and beautiful of the solids.

[Engraving, in André Thévet, *Les Vrais Pourtraits et Vie des hommes illustres . . .*, Paris, 1584. Paris, BNF. © BNF]

After ridding the world of the gods, the Greek philosophers and mathematicians imprisoned it inside a giant sphere, outside of which nothing could exist. This cosmic marble, of a rather modest size compared to that of the solar system, imposed itself as the image of the universe until the sixteenth century. Familiar, indestructible and impermeable, it was particularly reassuring.

[*Claudius Ptolemy at the Alexandria Observatory,* color woodcut, 1876. © AKG, Paris]

himself had the distinction of being able to hear this hypothetical music of the spheres, which remained inaudible to the common run of people. Finally, the Pythagoreans reportedly originated the term "cosmos," a sort of Great Everything within which harmony, beauty, symmetry and order reign.

PLATO'S UNIVERSE

The sphere became the definitive model for the universe through the work of the philosopher Plato (427–347 B.C.), who affirmed without batting an eyelid in *The Timaeus:* "Wherefore he made the world in the form of a globe, round as from a lathe, having its extremes in every direction equidistant from the center, the most perfect and the most like itself of all figures." The universe, whose center is occupied by the earth, is finite, unique and of an indeterminate size. It contains the totality of creation, and there are no worlds nor heavens beyond this one. But it is not enough to assert that the universe is shaped like a gigantic marble with our planet as its center to suddenly explain everything, such as the daily route of the sun in the sky, the cycle of the moon or the movements of the planets and stars. For Plato and his successors, we must also keep up appearances, that is, construct a cosmic edifice capable of reproducing, as faithfully as possible, the apparent course of the stars in the sky. All this without questioning the sphericity of the world or the circular shape of the paths followed by the heavenly bodies.

The model of the cosmos conceived in the fourth century B.C. by Plato's student Eudoxus of Cnidus is strikingly reminiscent of an onion. Each planet (the ancients only knew Mercury, Venus, Mars, Jupiter and Saturn), the sun and the moon were set within a sphere. These balls, of different sizes, were fitted one within the other. The last of these, the outer peel of the universe, contained the stars. The earth occupied the center of this system, with the other spheres rotating around it. They carried the stars along in their movement, which is why we see them move across the sky.

[From Andreas Cellarius, *Atlas coelestis seu harmonia macrocosmica,* engraving, seventeenth century. Stapleton Collection. © The Bridgeman Art Library]

the universe in the history of astronomy worthy of such a label. It comprised a set of hollow spheres (there were twenty-seven in all) fitted together like Russian nesting dolls. The outermost one, which enveloped the universe, was known as the sphere of fixed stars, because the stars were pinned to it like butterflies. The sphere spun from east to west around its axis, and carried the stars along in its movement. The planets, sun and moon were each set within their own sphere. To reproduce their trajectories, which are more complex than those of the stars, Eudoxus attached them to two, or even three other spheres. They were animated by different movements and arranged in such a manner that they carried each other along mutually, so that the combination of their rotations matched their observed journeys in the sky. Finally, the earth sat perfectly immobile at the center of this clever framework. Once this machinery was set in motion it reproduced, roughly, the overall movement of the celestial bodies as seen from the earth. This is what is called a model of the universe: an attempt to reconstruct celestial reality, a way of keeping up appearances that may have nothing in common with the true mechanisms at work in nature. What matters is that it works.

3.

THE SPHERE OF FIXED STARS

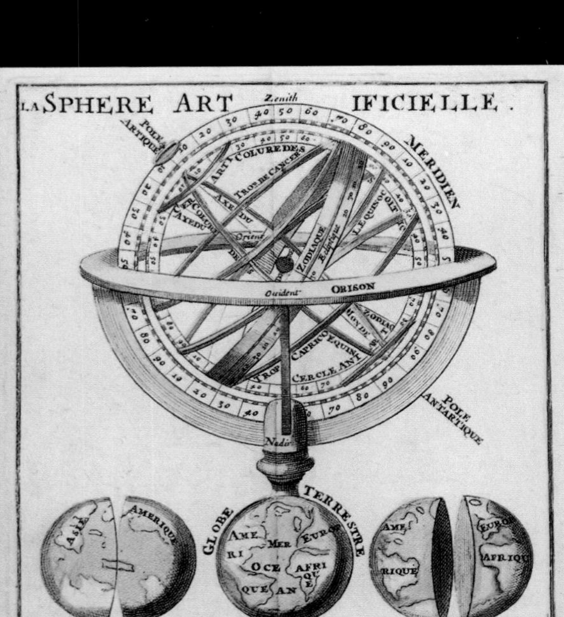

Armillary spheres, now icons of astronomy, are simplified representations of the celestial sphere. They help us visualize the movement of the earth above a horizon.

[*The Artificial Sphere,* color copper engraving, France, seventeenth century. Paris, private collection. © AKG, Paris]

The model of the universe underwent constant retooling until the Greek geographer, mathematician and astronomer Claudius Ptolemy put the finishing touches on it in the second century A.D. His ingenious spherical clockwork, with its intertwined circles and movements, marked the apex of astronomical knowledge until the end of the sixteenth century, when problems of planetary motion became so troublesome that the model had to be tossed into the junk drawer of science. For reasons of a practical order, however, modern astronomers have retained one of the principal elements of this subtle cosmic framework: the so-called sphere of fixed stars.

OBSERVATION OF THE STARS

The hollow marble that supposedly houses the stars in Eudoxus's system was not a complete fabrication of the Platonist's fertile mind: beyond the purely aesthetic and philosophical criteria that reigned at the time of its birth, it was also the child of observation. The particularities of the movement of the stars, visible to any observer in the Northern Hemisphere with nothing other than a good pair of eyes and a bit of time, all support this model.

The Horizon

Whatever one's level of erudition and culture, the person who decides to spend several nights contemplating the sky inevitably returns to the Cro-Magnon state. If by chance we set aside everything we have learned and

up against this boundary, the limit at which our gaze leaves the earth. This is why, wherever we are, we always have the impression of standing at the center of a flat, rigid disk of an infinite diameter—the horizon.

The Tilt of the Horizon

Keeping in mind that the earth is round, the plateau of the horizon closely resembles a dish resting upon a large balloon. When an observer is situated right at the North or South Pole, the dish sits perfectly horizontally: it is parallel to the equator and perpendicular to the earth's axis. This same pancake placed upon the equator becomes vertical, parallel to the earth's axis. Between these two extreme positions, the horizon will be more or less tilted depending on where the observer is located.

THE HORIZON

The earth is round. On a day-to-day basis, however, the horizon causes us to experience our planet as a flat pancake. Our gaze cannot grasp the shape of the earth: the horizon runs straight across our field of vision, following a perfectly rectilinear and horizontal trajectory. Objects resting on the earth's surface—trees, houses, mountains—found along this route are visible to us. Up to a certain point, that is, beyond which it is impossible to know what is happening on our planet, even with the most high-powered binoculars. Our line of sight disappears into infinity, as it is unable to bend and follow the curvature of the earth. In any direction we look—east, south, west, north—we systematically run

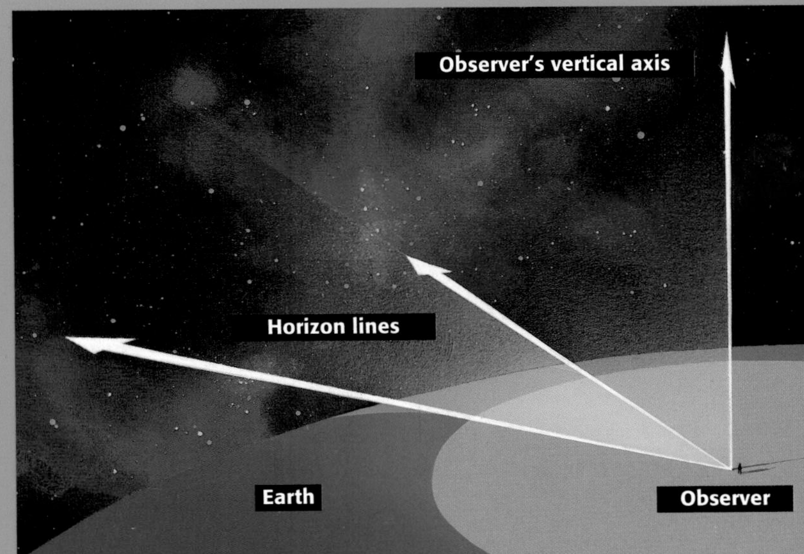

agree to grasp the world using our five senses alone, we will leave with precisely the same impressions our ancestors did. The earth, for example, will appear flat to us. Unable to see the earth's curvature, we have the impression, wherever we stand, of being at the center of a circular plateau of infinite dimension, the horizon (see sidebar, p. 23). Only objects found on this flat dish or above it are accessible to us. The rest of the world, buried beneath the horizon, remains invisible.

The Fixed Stars

To an observer, the stars all seem equally distant from the earth. To the naked eye, there is no way of grasping the depth of the sky, and it is impossible to say whether one star is further away than another. The stars look like little points, more or less brilliant, more or less colored, set upon a uniformly black background. They remain very still. However untrained an observer

According to the Greek myth to which it owes its name, Ursa Major (literally "The Great Bear," a.k.a. the Big Dipper) was none other than the beautiful Callisto. Zeus became infatuated with this nymph of Artemis, and with her produced little Arcas—much to the annoyance of Hera, his jealous wife, who transformed Callisto into a bear. The poor beast wandered for years in the forest, until, one day, she came face-to-face with her son. Wishing to hug Arcas in her hairy paws, she nearly perished upon Arcas's sharp spear. Zeus rescued her in the nick of time by transporting the bear into the sky.

[© *Ciel & Espace*/J. Lodriguss]

we may be, we will have no trouble locating the constellation Ursa Major (literally, "The Great Bear," a.k.a. the Big Dipper) amidst the starry jumble: it is shaped like a ladle. Four stars suggest the utensil's handle, to which is attached a square sketched out by four other celestial dots. We can return to contemplate it night after night, for years if we fancy it, yet Ursa Major will always retain its utterly recognizable silhouette. None of the eight stars that provide its so characteristic appearance will budge during our entire lifetime—we will never catch them wandering around. To us, the stars are fixed, like butterflies pinned for eternity upon the screen of the sky.

The Movement of the Stars

And yet, they move . . . They travel from east to west, their path shaped in an arc. Soon after nightfall, we the observers, at a slight advantage over the Cro-Magnons because we possess a compass, find ourselves facing south. On our left, that is, toward the east, we see a handful of stars rise above the horizon. If we pay close attention, we can identify within this cloud of white pinheads at least one constellation, which will serve as our point of reference. With the passing hours, it will slowly but surely ascend the firmament. Then its route will bend, and it will begin its slow descent toward the western edge of the horizon, beneath which it will gradually disappear (see sidebar, p. 27). Strange . . . We come back to the same spot the very next evening, at exactly the same time. And we observe, once more, our constellation rise in the east to begin the same journey

THE SPHERE OF FIXED STARS

In the theoretical model of the sphere of fixed stars, the sky is a hollow sphere to the inside of which the stars are pinned. The observer sits at the center of this infinitely sized marble, and the sphere of fixed stars rotates from east to west. Its axis of rotation passes through its north and south poles, extensions of the earth's own poles. The stars, dragged along by the sphere's movement, draw circles around the earth.

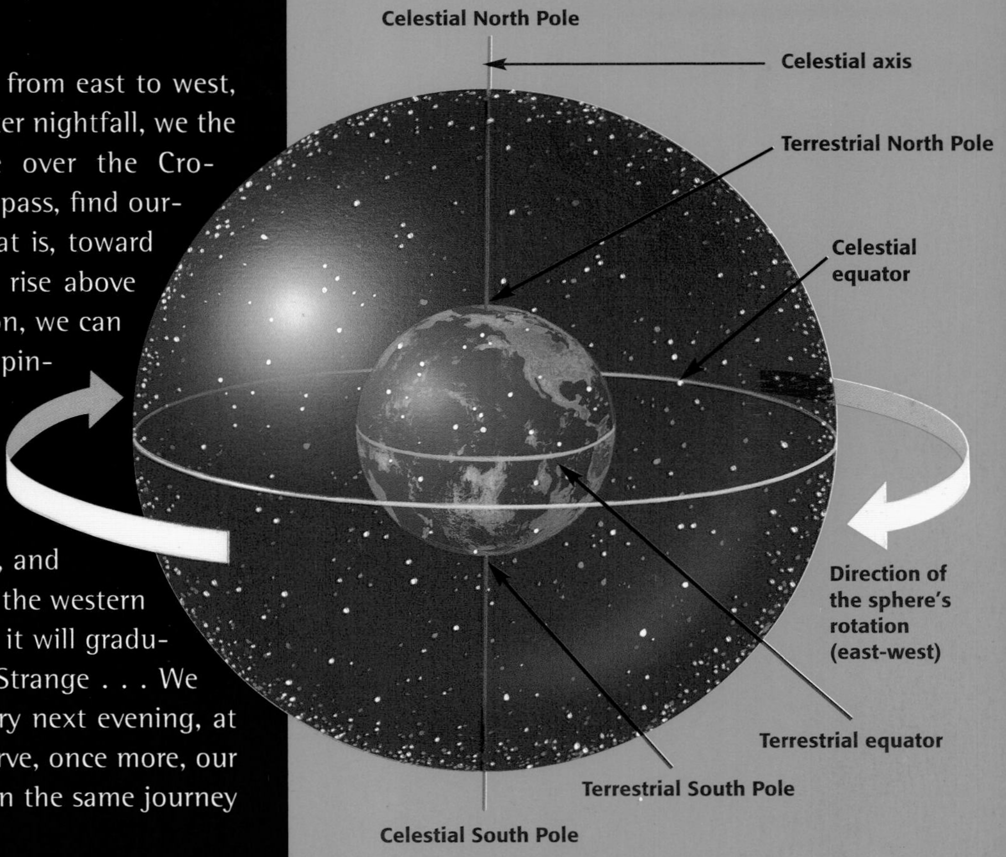

Celestial North Pole
Celestial axis
Terrestrial North Pole
Celestial equator
Direction of the sphere's rotation (east-west)
Terrestrial equator
Terrestrial South Pole
Celestial South Pole

as the evening before. The phenomenon repeats itself over the ensuing nights, and we notice that it comes back regularly, about every twenty-four hours, to occupy, with little variation, the same position in the sky as the evening before. The conclusion is logical: the constellation is revolving around us. Its arc-shaped course, its nightly return and faithfulness to our nocturnal rendezvous all lead us to this deduction. Although we have access only to the part of its journey that takes place above the horizon, it is almost certain that this journey continues below.

A Conveyor Belt

Absolutely all the stars in the sky (except for one) move. During the climb and descent of "our" constellation, we realize that new stars continually appear to the east, while scores of others disappear to the west. From the moment our reference star appears to the time it is swallowed up, the sky completely changes its appearance. Stars that were below the horizon have now risen above it, and vice versa. The parade of stars seems uninterrupted. Everything moves in the same direction, following circular arcs, larger or smaller depending on their proximity to the southern edge of the horizon (see sidebar, p. 27). In short, the stars are all spinning around us, but without any one of them moving a comet's hair away from the position it holds in relation to the others. They move together, in one sole block, exactly as if they were attached to a monstrous conveyor belt above our heads. Conclusion: it is not the stars that move, but the wall to which they are screwed. The sky, that is . . .

THE EARTH'S SKIES

Playing with the model of the sphere of fixed stars, the size of the earth is negligible in comparison to that of the surrounding sphere. We can act as if the plateau of the horizon were our planet; since we appear to occupy its central point, we can consider ourselves at the very center of the celestial sphere. Now, the horizon stretches out toward infinity, dividing the sphere of fixed stars in two halves. We can only see what is located on the horizon or above it: only one half of the starry marble is accessible to us, the aptly named celestial vault, that covers us like a hat.

The sky seen from the North Pole

N

Apparent trajectories of the stars

1. The horizon of an inhabitant of the North Pole cuts the sphere of fixed stars right at the level of the equator and is perfectly perpendicular to the axis of rotation. The North Pole of the celestial marble is located directly above the observer, who will never see the stars rise or set: they trace their circular paths parallel to the horizon, and their paths appear shorter the closer they are to the pole. The observer will never have access to the stars of the Southern Hemisphere of the celestial sphere, which remain hidden beneath the horizon.

2. At the equator, the plane of the horizon cuts the sphere in the north-south direction. It is parallel to its axis of rotation, and the stars trace perfect semicircles vertical to the observer. Dragged along by the sphere,

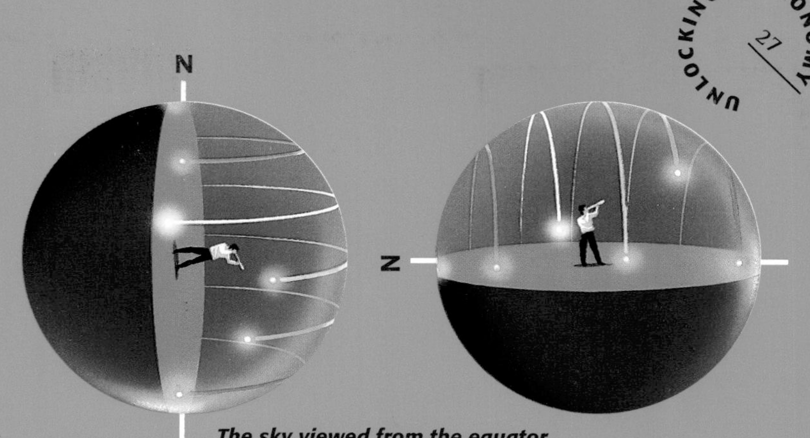

The sky viewed from the equator

stars located beneath the horizon will emerge one after another from the east, permitting the inhabitant of the equator to admire the total contents of the celestial sphere except for the North Star.

3. The horizon of an observer placed at any other location in the Northern Hemisphere cuts the celestial sphere at an angle. It is tilted in relation to the axis of rotation, and the North Pole of the celestial sphere, occupied by the well-named North Star, is easily visible in the sky. Some stars, the circumpolar stars, turn full circle in the sky and remain visible all night long. The others can be seen only when they pass above the horizon, and their trajectories, of varying length, resemble tilted arcs of a circle. Finally, a small number of stars of the Southern Hemisphere are accessible to the observer.

The sky viewed from a median latitude in the Northern Hemisphere

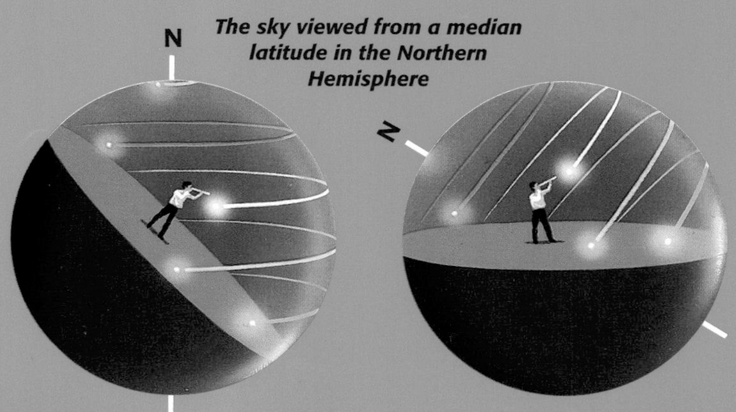

THE CELESTIAL SPHERE

Everything makes sense if the observer accepts that the sky is shaped like an enormous hollow sphere, inside of which the stars are fixed. The earth remains perfectly still at the center of this marble, which rotates around it in the east-west direction. It takes about twenty-four hours to complete a revolution. Dragged along by its movement, the stars spin around the earth. The observer can only see the portion of their trajectory that takes place above the horizon.

Observed from the earth, the stars all seem in movement. In the Northern Hemisphere, they trace circles around a very particular point in the sky, occupied by the North Star, a star in the constellation Ursa Major.
[© *Ciel & Espace*/P. Parviainen]

A Sky in Rotation

Additional evidence to support this hypothesis: the circumpolar stars. These are the stars that remain visible throughout the night, tracing circles around a point above the northern edge of the horizon. The further away from this point a star is located, the greater the diameter of its circular path. This minuscule pivot point, today occupied by a star called the North Star, represents the North Pole of the celestial sphere. Let us imagine a rod running through this point. It would penetrate through the North Star, cross the observer's horizon and reemerge through the South Pole, located directly opposite that of the North (see sidebar, p. 25). Much like a chicken roasting on a spit, the celestial marble rotates around this axis. Forced to follow its movement, the stars turn around it, too. But the closer they are to the two poles of the sphere, the smaller the circles they draw. The poles themselves remain perfectly immobile, and the North Star is the only star in the sky that does not drift toward the west.

ınd yet, it moves . . .

ike good Cro-Magnons, if we trust solely our observa-ons and impressions, we will assert that the sky evolves around the earth. But like good nineteenth-entury scientists, we know full well that this is false: ne sky does not move. Rather, it is the earth that spins om west to east around its axis. It drags us along in its ıovement, and it is we who move—toward the east—in elation to the stars. The star's nightly rounds are ıerely an illusion. According to the so-called corkscrew rinciple (see adjacent sidebar), a spherical sky spinning om east to west around an immobile earth, or an arth in rotation from west to east in a space frozen nd dotted with white points amount to exactly the ıme thing: observers will have the impression in both ıses that the stars are spinning around them. The two erspectives are completely equivalent and, whether we dopt one or the other, the apparent length of a celes-al body's journey and the time it takes to return to a ven position in the sky will be identical. Under these onditions, the astronomer, a practical animal, has tained the sphere of fixed stars as a working tool. ecause it conforms to observations and very neatly ccounts for the apparent shifting of the stars, stronomers have preserved it as a theoretical model. A ctitious and purely geometric representation of the niverse, it has the enormous benefit of letting us ssign precise positions to the stars and to follow and udy their movements. To consider oneself the center f the universe sometimes has its advantages.

THE CORKSCREW PRINCIPLE

Uncorking a bottle is, in principle, a rather simple exercise. You begin by sinking a corkscrew into the little cylinder object that obstructs it—that is, the cork. It is possible to go about this in two different ways: either by turning the corkscrew clockwise while keeping the bottle still, or by turning the bottle counterclockwise while keeping the cork-screw still. The two methods have exactly the same effect. In both cases, the corkscrew sinks into the cork; all you have to do is pull to uncork the bottle. In theory, when con-fronted with an open bottle it is impossible to know which one of the two, the corkscrew or the bottle, remained still and which one moved. In reality, only one method, the first, is used. But what differ-ence does it make, given that the second operation, the mirror opposite of the first and thoroughly imaginary, yields precisely the same result?

2 Guides to the Sky

1. THE CONSTELLATIONS 2. THE ROUTE OF THE SUN

Map of the Northern Hemisphere of the sky by Mr. de La Hire, royal professor and member of the Academy of Sciences (1775). On the bottom right, a little *Fleur de Lys* constellation is visible between Aries and Perseus. A fabrication in homage to the King of France, it has disappeared from maps.
[© *Ciel & Espace*]

The sphere of fixed stars literally froze t
making it a sort of ethereal wall on wh
stars are solidly fixed, still as pictures hanging or
They appear to be in motion, by no effort of the
and always occupy the same position on the in
the celestial marble, which merely spins rou
round. This vision of the universe is at the origin
of astronomy's craziest projects: to count the
population of the sky and copy it down as faith
possible while respecting the proportions and di
separating the stars in the gigantic, spangled fres
runs along the walls of the universe. A sky map
freehand sketch. Each star's position and its loca
the sphere must be transferred to its correspondi
in three dimensions—a miniature celestial gl
onto a sheet of paper. For this, the astronomer
tool that is indispensable to all cartographers: a

of reference. Such a system consists of a set of reference points, markers that allow an object to be assigned a precise place by indicating the distance separating it from all the others. A good sport, the celestial sphere provides a whole heap of such indicators: basic planes like the horizon, equator and celestial meridians, or even the ecliptic, the annual route of the sun. This toolbox has proven so practical that it is still used today to pinpoint the positions of the stars.

1.

THE CONSTELLATIONS

Before the invention of the sphere of fixed stars, the only method our distant ancestors had to locate stars came from their powerful imaginations: they envisioned the constellations.

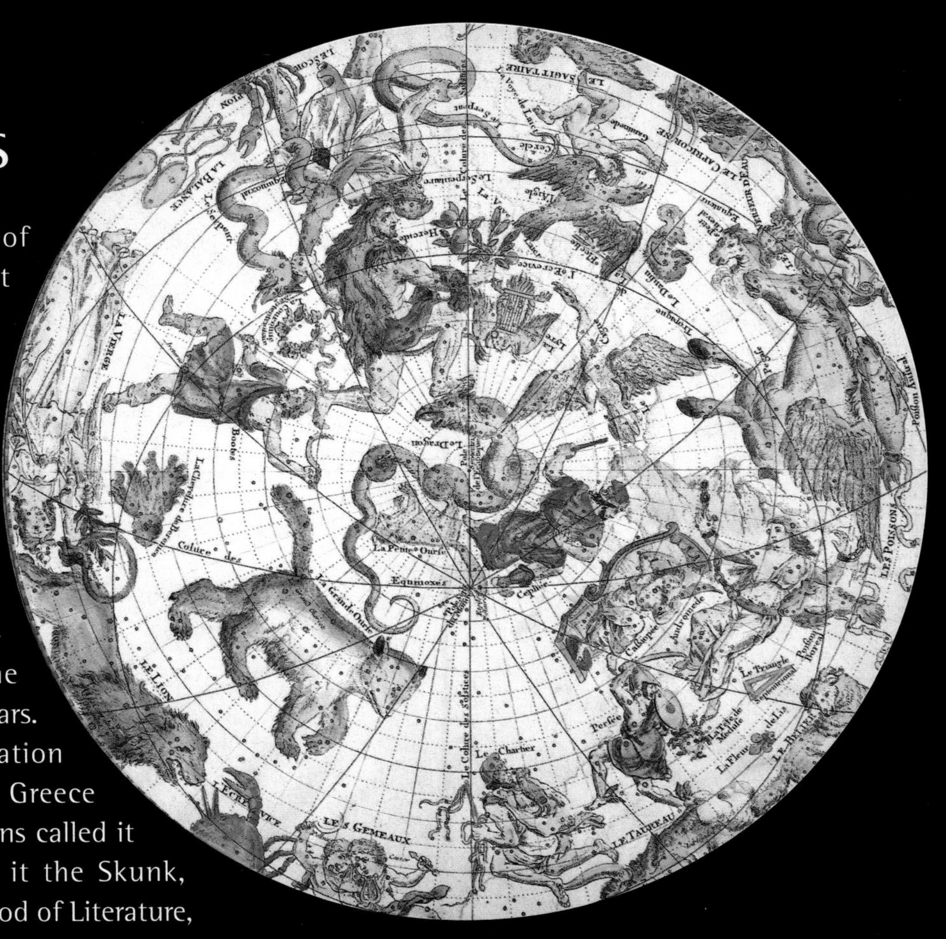

THE WORLD AND ITS MANY SKIES

Totally imaginary, inspired by mythology, the animal world and everyday objects, every civilization has interpreted the figures suggested by the particular arrangement of groups of stars. The Egyptians thought the constellation seen as a Great Bear on the isles of Greece resembled an ox leg. The Mesopotamians called it the Chariot, the Lakota-Sioux called it the Skunk, while the Chinese saw Wen-chang, the god of Literature, receiving his Minister of World Literary Affairs and divine escort. To situate a star, the ancients needed only indicate its position within the constellation, taking care to note which vision of the heavens they had in mind. For some, the star Alkaid, for example, is found at the end of the Great Bear's tail, while for others, it represents the ox's hoof or Kuan-ti, the god of War, third and final member of the well-read Chinese minister's entourage.

Olympus in the Firmament

The diffusion of Greek astronomical knowledge across the Arab and Western worlds eventually established Hellenic mythology's rowdy pack of goddesses, gods and heroes as the ultimate system of reference. The celestial vault resonates with the distinct echo of their many adventures, to the great joy of authors of celestial atlases who gave free rein to their imagination and peopled their maps, supposedly faithful representations of the sky, with teats, thighs, buttocks and Herculean torsos. One of the most disturbing figures in this mythical sky is the constellation of Scorpio. On the orders of the goddess Artemis, Scorpio now spends eternity pursuing the famous hunter Orion. The latter, usually depicted as a robust fellow brandishing a club, made the mistake of preferring the beautiful Aurora over the divine huntress, who sent the scorpion to avenge her with its deadly sting. He's got a long way to go: Orion sets at the very moment Scorpio rises. The great silhouette of the nasty arthropod is sketched out by a handful of stars, including the magnificent Antares. Flaming like a ruby, Antares is found on Scorpio's back and represents the creature's burning heart.

Scorpio never stops moving along with the rotation of the celestial sphere, and you may have to hunt around the sky for a minute to catch a glimpse of him. Trying to locate Antares this way is roughly reminiscent of flying over the Atlantic Ocean for hours in the hope of seeing the passengers on board a cruise ship floating somewhere between Southampton and New York.

The Geographic Coordinates

Earthlings have always been able to know exactly where they are. Whether smack in the middle of the Gobi Desert or Pacific Ocean, they have at their disposal a kind of toolbox, the system of geographic coordinates, to keep them from ever getting lost (see sidebar, p. 35). The planet has been marked out in a dense grid of crisscrossing imaginary lines, the meridians and parallels. The former run vertically and pass through the earth's two poles. The latter run horizontally, parallel to the equator. Any object on the earth's surface is at the intersection of two of these lines. The meridian on which the object sits is located according to the distance—called longitude—that separates it from a meridian chosen as a source: the one that runs through the English observatory in Greenwich. The parallel is determined by the position—the latitude—the object occupies in relation to the earth's equator. With the aid of this system, sailors and explorers establish their precise points of departure and arrival, trace their route and confirm at regular intervals that they are not veering off course.

Ninth-century Arabs acquired the catalogs and manuals of ancient Greek astronomy, in which the stars' positions and descriptions of their movements were given within a well-organized sky. Rather than starting from scratch, Arab astronomers adopted the celestial imagery of the Greeks.

[Map of the sky in two hemispheres by Khalifa[...] Paris, BNF. © BNF]

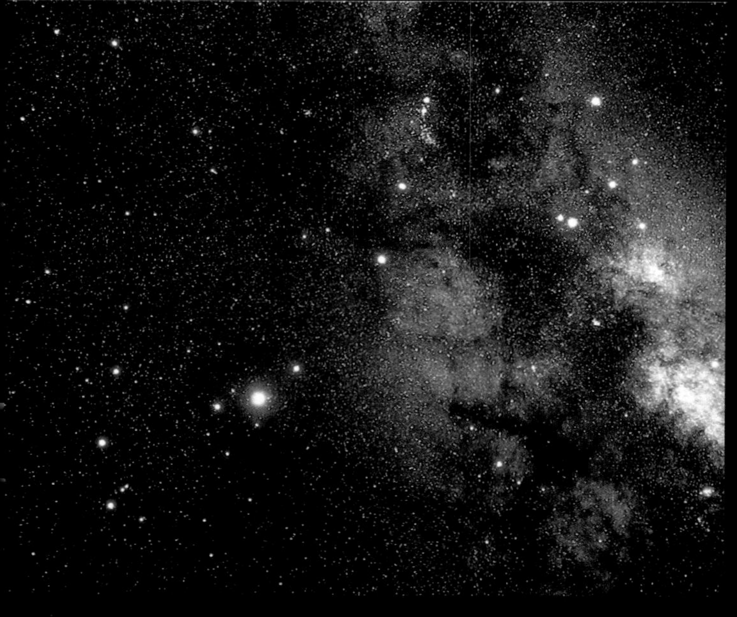

The constellation Scorpio, visible in the summer, is fairly easy to locate because of its principal star, the red Antares, that sits on the creature's back. The star's name is derived from Anti-Ares, the rival of Ares—that is, of Mars, the red planet. [© *Ciel & Espace*/A. Fujii]

The Horizontal Coordinates

The astronomer's resources are not as limited as they might seem. What about the horizon? This flat plate at whose center we sit always serves as our first point of reference; it is in relation to the horizon that we observe the movement of the stars. Why not use this plane to assign a precise position to Scorpio's Antares? We need only treat the horizon as the equator of a giant, virtual ball with the observer at its center, and whose walls contain the sky. We can then lay a grid upon this marble just as we do with the earth, with a network of lines parallel to the celestial horizon and meridians. Antares, like a human explorer, can always be found at the intersection of two of these lines. In principle, you need only figure out which two to determine the location of any star in the sky. It's easy: the meridian on which the star is located is determined by its distance—named azimuth—to an arbitrary meridian, which passes through a precise point on the horizon. The parallel is characterized by its height in relation to the horizon. This system of coordinates, called horizontals, allows us to study and track the movement of a star (see sidebar, p. 37). But the location of a star, calculated according to this method, is constantly changing. The reference point provided by the horizon remains immobile only to the observer. The celestial sphere is rotating in relation to us, and carries Antares along in its movement. The star never stops moving across the sky, leaping from one celestial meridian and parallel to the next. Nor does the horizon have the same tilt, depending on where the observer is standing: the star's height, measured at the same moment in Alexandria or Copenhagen, will be very different.

THE GEOGRAPHIC COORDINATES

The surface of our planet has been carved up into a network of circles. The first set slices it from pole to pole in rounds parallel to the equator: these are the parallels. The second set, the meridians, pass through the two poles and are perpendicular to the parallels. They cut the earth up into quarters (or zones) like an orange. Wherever we happen to find ourselves, we are inevitably standing at the intersection of a parallel and a meridian. We need only know which ones to define our position precisely. The parallel can be determined using the latitude: it is the value of the angle made by the straight line connecting the edge of this circle to the center of the earth and the plane of the equator. The latitude is measured in degrees, from 0 to 90, from the equator out to the poles (the latitude of the former is valued at 0 degrees, with that of the latter at 90 degrees). You must specify whether this is north or south, depending on what hemisphere you're in. The meridian is determined by its longitude; this corresponds to the angle, measured at the equator, made by the meridian in question with that of the meridian of reference, the one passing through the English observatory in the city of Greenwich. The longitude is counted from 0 degrees, the value of the Greenwich meridian, to 180 degrees. The measurement can be done in both directions, east or west, in relation to Greenwich.

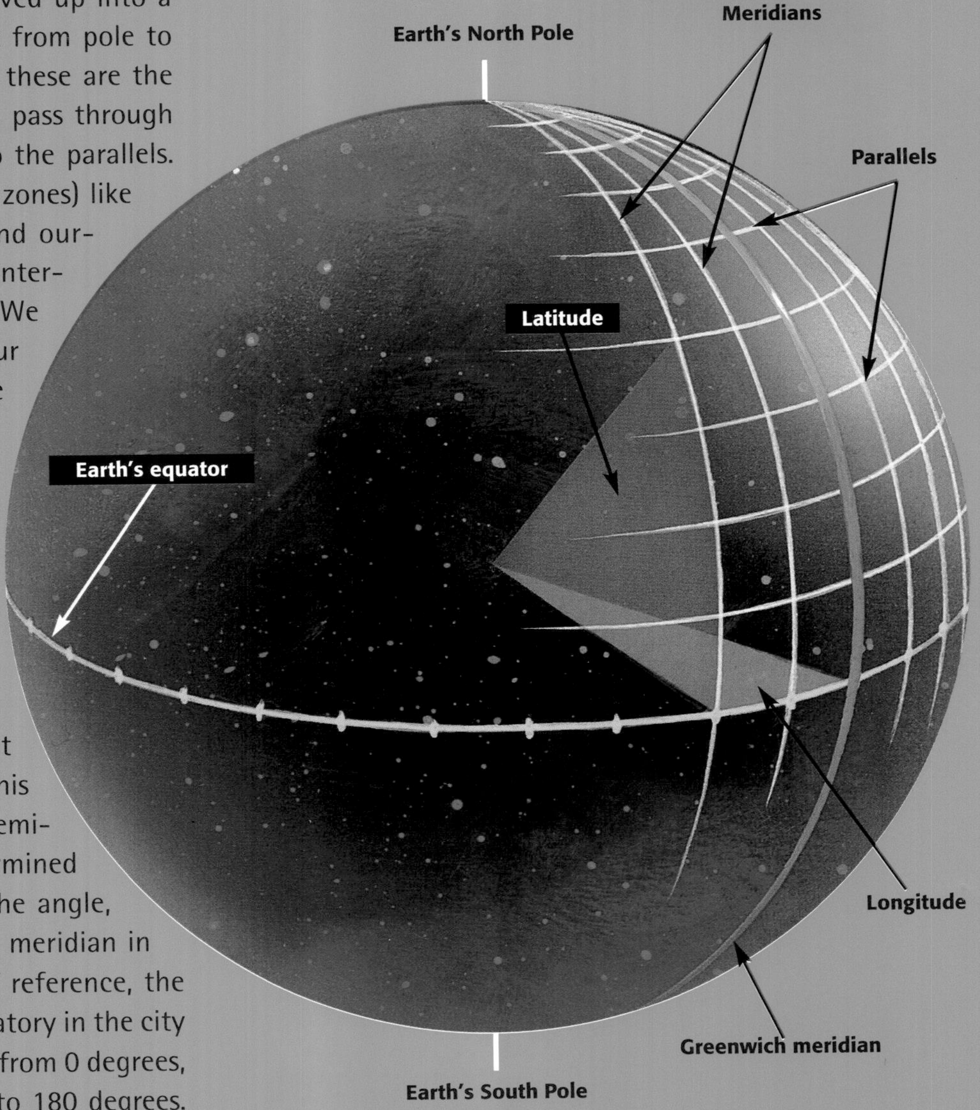

The earth, divided into squares by a system of meridians and parallels.

[World map by J. Hondius, after G. Mercator, engraving, 1630. © AKG, Paris]

2.

THE ROUTE OF THE SUN

The stars, hanging from the same nail on the rounded walls of the cosmos, can be compared to a tree firmly rooted in the soil of the earthly sphere. On our planet, the position of any particular object can be determined in relation to the Greenwich meridian line and to the circle of the equator. Neither moves; they remain fixed within the earthly anatomy like hoops frozen in a block of ice. Their distance to the tree does not vary one iota, however the earth might move. In the system of geographic coordinates, the location of immobile objects—an island, a coast, a city—is unique, valid for any earthling, in any spot on the globe and any hour of the day. To be able to assign a fixed position to a star, you must find two points of reference that are an integral part of the celestial sphere and do not move along with it. The celestial equator, an extension of that of the earth, takes care of this problem, and all that remains is to discover a celestial equivalent of Greenwich as an immobile point of reference in order to assign the stars their definitive coordinates. This reference point is kindly provided by the sun.

THE HORIZONTAL COORDINATES

Whatever may be the intersection of earthly meridian and parallel on which we stand, we will always have the impression of being at the center of a vast plateau, the horizon, which cuts the celestial sphere into two equal halves. Lifting our eyes to the sky, we have the distinct feeling of being covered by a hemispheric dome dotted with stars, whose summit is located directly above our heads. Everything behaves as if we were standing at the center of our own celestial sphere, with the horizon as its equator and its North Pole, called the zenith point, situated just above our head. The South Pole of this sphere is the nadir, opposite to the zenith. To assign a position to a star, the observer begins by determining the celestial equivalent of the latitude, that is, the height of the star: this is the angle made by the straight line connecting the star to the observer with the plane of the horizon. It is measured from 0 degrees, the height of the horizon, to 90 degrees, the height of the zenith. Then, the observer finds the azimuth of the star: that is, the angle, measured upon the horizon, made by the line passing through the zenith and the star with a second line of reference. For the majority of astronomers, the latter is the one passing through the zenith and point S, which represents due south (surveyors prefer using due north). The azimuth is counted clockwise starting from this plane, from 0 to 360 degrees.

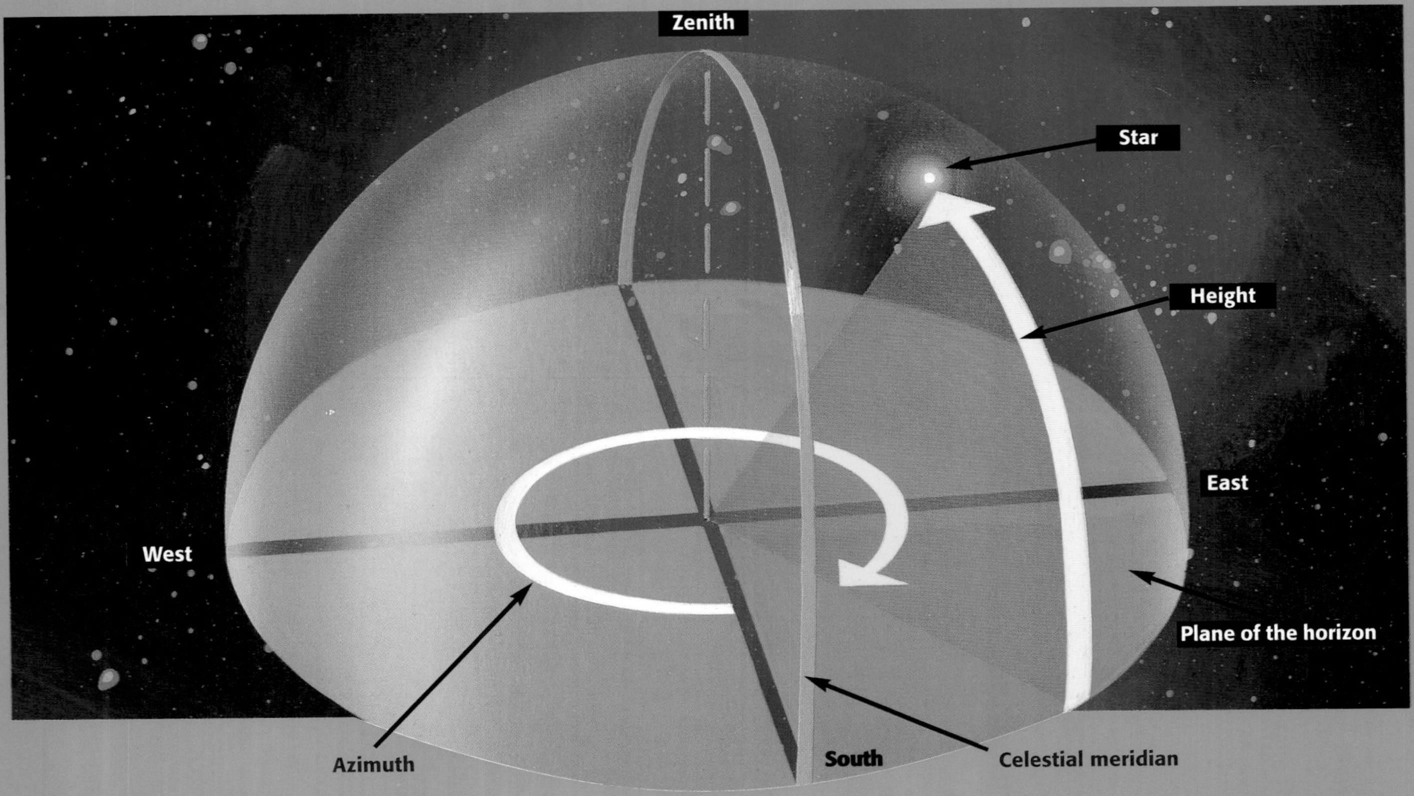

Zenith
Star
Height
East
West
Plane of the horizon
Azimuth
South
Celestial meridian

THE APPARENT MOVEMENTS OF THE SUN

The corkscrew principle has allowed astronomers to pin down the sun onto the sphere of fixed stars . . .

Alternating Day and Night

Our planet spins on its axis before an immense lamp, the sun, which we can imagine as unmoving. The half of the earth that faces this solar lightbulb is inundated with light, while the other side is plunged in darkness. Dragged along by the movement of the earth, the dark territories enter, one by one, into the lit zone, while the illuminated regions are, bit by bit, pushed into night. Taking into account the direction of the earthly rotation from west to east, the entry into the light is made in the east and the departure takes place in the west. This is why when day rises for people in North America, those in China are ready to be tucked into bed.

Diurnal Movement

Despite everything we know, we fall systematically victim to the illusion that gave birth to the celestial sphere. We will always have the impression that it is the sun that moves, rather than the earth. At the instant when, carried on the earth's back, we enter the lit zone, we see the sun appear in the east. It rises above the horizon and follows a certain path across the sky while, in reality, we are the ones actually moving through the sunlight. The moment we return to darkness, the sun sets in the west and disappears beneath the horizon. This apparent movement closely resembles that of the stars: just like them, the sun seems to spin around the earth. From this point, it takes only one logical step to conclude that the sun, too, is fixed to the interior of the celestial sphere and is forced to follow its movement.

Annual Movement

Spinning around on its own axis, the earth also revolves around the sun: in a year, it traces a circle-like orbit around the daystar. Our planet's axis, the imaginary rod joining its North and South Poles, is not perfectly perpendicular to this loop: it is tilted in relation to its vertical by 23.44 degrees. The earth does not stand straight in its orbit, but leans slightly. The seasons are the result of both this tilt and the annual movement of our planet. Because of these two factors, the earth changes its position in relation to the sun throughout the year. Around 21 June, for example, it leans toward the sun with the Northern Hemisphere facing forward: this is the beginning of summer. Six months later, when the earth is leaning backward and pushes forth its Southern Hemisphere, the northern one enters winter (see sidebar, p. 41).

To the common observer, oblivious to the movement of the earth and the tilt of its axis, it is the sun that moves rather than our planet. In summer, when the earth tips forward, we see the daystar climb northward. In winter, when the planet leans backward, we see the sun descend toward the south. This movement is only apparent, but it allows us to behave as if the earth were immobile and straight on its axis. To reproduce the cycle of the seasons precisely, we need only have the sun follow a somewhat peculiar path around our planet. This fictitious orbit, the ecliptic, is a circle tilted in relation to the earth's equator at an angle of 23.44 degrees: the sun takes a year to complete it, climbing northward in summer and tumbling southward in winter (see sidebar, p. 42). Careful: the ecliptic also refers to earth's orbit around the sun. It all depends on your point of view . . .

Armillary spheres were first and foremost pedagogic tools, scale models of the universe used to highlight its major points of reference. First among these reference points is the circle of the horizon (in pink), in relation to which the sphere's axis of rotation is tilted. The yellow circle represents the meridian, while the thick hoop of the ecliptic, with its zodiacal compartments, slices the celestial equator at an angle.

[Mustafa ibn Abdullah, Turkish armillary sphere, from *Le Livre de la description du monde*, Paris, BNF. © BNF]

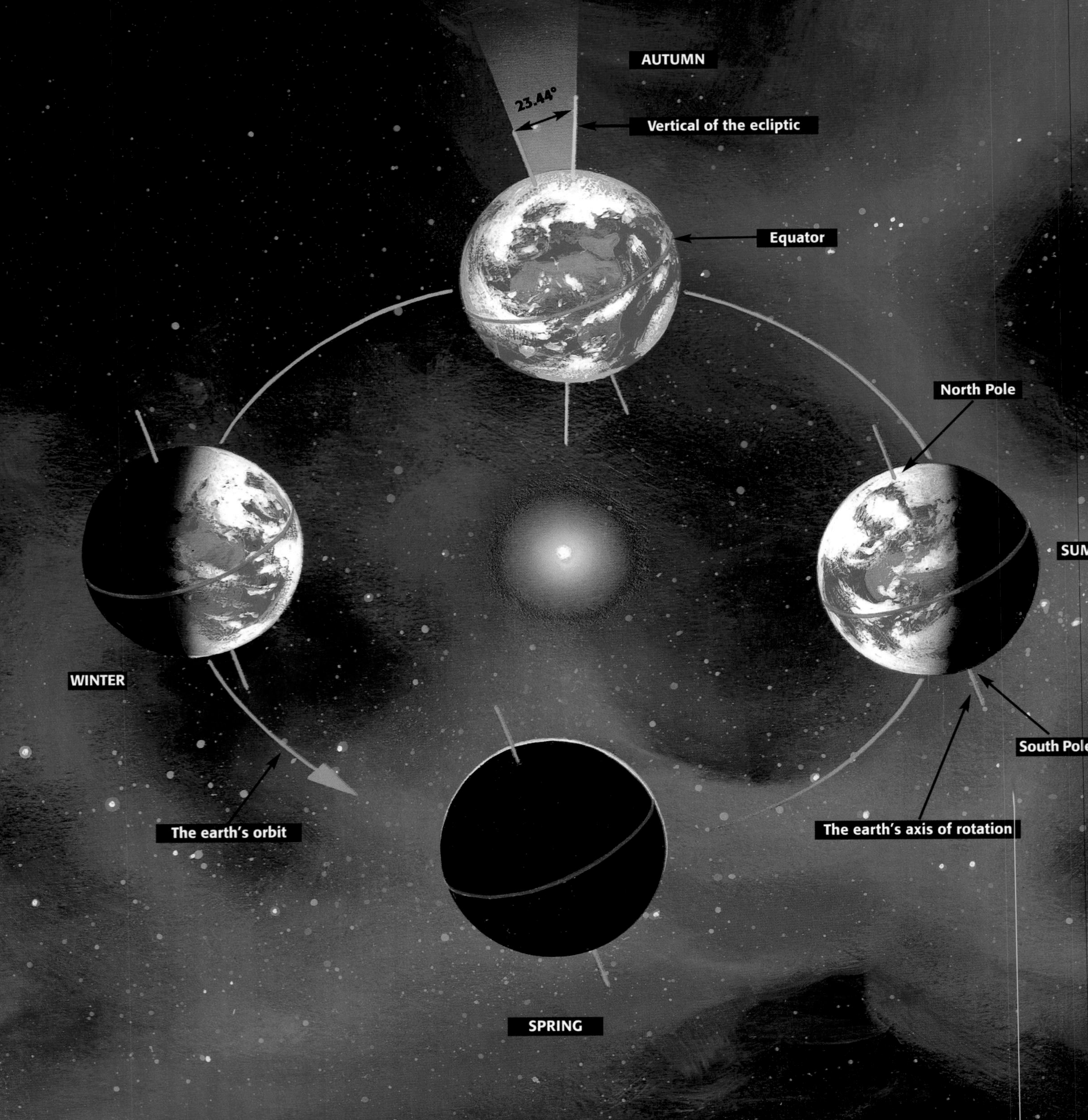
AUTUMN
23.44°
Vertical of the ecliptic
Equator
North Pole
WINTER
SUM
South Pole
The earth's orbit
The earth's axis of rotation
SPRING

THE FOUR SEASONS

The earth revolves around the sun following a path shaped like an ellipse—in fact, a slightly flattened circle whose diameter measures an average of 93.2 million miles. It takes a year to return to its starting point.

Our planet's axis, the imaginary rod that passes through its North and South Poles, is not perpendicular to this path: it is tilted in relation to the vertical of the ecliptic at an angle measured today at 23.44 degrees. This value and orientation of the axis do not change during the journey the earth takes around the sun: the earth always leans to the same side, wherever it may be in its orbit.

At the moment of the summer solstice, which occurs around 21 June, our planet is leaning directly toward the sun. We are tilted toward it, the Northern Hemisphere leaning forward and the Southern Hemisphere leaning backward. The plane of the ecliptic cuts the great terrestrial body above the equator at the level of the tropic of Cancer, onto which the solar rays fall vertically. The tilt of these rays is far greater in the southern hemisphere, where it's rather chilly outside, than in the Northern Hemisphere, which is getting ready for the summer heat.

Three months later, the earth has moved. It has changed position in relation to the sun and no longer leans toward it, but to the side. Its two hemispheres share the solar light equally, which falls vertically upon the equator. The north is cooling down, autumn is beginning to lose its leaves, while the south is finally emerging from winter. It is around about 22 September, the date of the autumnal equinox.

The planet continues its route and, around 21 December, occupies the diametrically opposite position in its orbit to the one it held on 21 June. It offers the sun its southern pole, and the solar rays fall vertically onto the tropic of Capricorn, below the equator. It's the Southern Hemisphere's turn to enjoy the delights of summer, while the north slowly sinks into winter.

Until 21 March, that is, the date of the spring equinox. The earth, tilted toward one side, looks the same as it did at the moment of the autumnal equinox. The Northern and Southern Hemispheres have become equal again in relation to the sun. The former warms up and buds, while the latter breaks out the fall wool. Three months later, it's the summer solstice again, and the cycle of the seasons starts over.

THE SUN'S ROUNDS

It is possible to reproduce the succession of seasons by imagining the earth still and straight on its axis. You need only behave as if the sun itself revolved around us in a west-east direction. The plane of its orbit is tilted in relation to the earth's equator at an angle measuring 23.44 degrees. The duration of this fictitious revolution of the sun is the same as that of the earth, or one year. The summer solstice would thus occur the moment the star reaches the highest point in its trajectory in relation to the earth's equator. It sits vertically above the tropic of Cancer and floods the Northern Hemisphere with its light and heat, while the Southern Hemisphere shivers. Three months later, it has descended southward and occupies the point at which its orbit crosses the plane of the earth's equator. It shares its light and heat equally between the two halves of the earth; this is the autumnal equinox. It continues its route until the day it reaches the lowest point in its orbit. This is the winter solstice: the sun sits just above the tropic of Capricorn and deprives the Northern Hemisphere of heat. From this point on, it can only begin climbing back up and finds itself at its orbit's second point of intersection with the plane of the earth's equator, which astronomers have named the vernal point: this is the spring equinox. Three months later, it's the summer solstice again.

THE ZODIAC

The sun strolls amidst the stars. It crosses several constellations, which mark out so many milestones along its annual journey. The stellar belt that encircles the ecliptic was named "zodiac," the circle of living beings, by the Greeks. Our forebears divided it into twelve sections of equal length—the signs. Each housed a particular constellation: Pisces, Aries, Taurus, Gemini, Cancer, Leo, Virgo, Libra, Scorpio, Sagittarius, Capricorn and Aquarius. This celestial highway traditionally represented as geometric, as in the facing diagram, is a bit too good to be true. The actual boundaries of the constellations crossed by the sun do not coincide at all with the boxes in which the ancients sought to contain them. Each of the figures of the zodiac occupies a very variable surface on the celestial vault and a thirteenth constellation, Ophiucus (not represented in the illustration), has slipped in between Sagittarius and Scorpio.

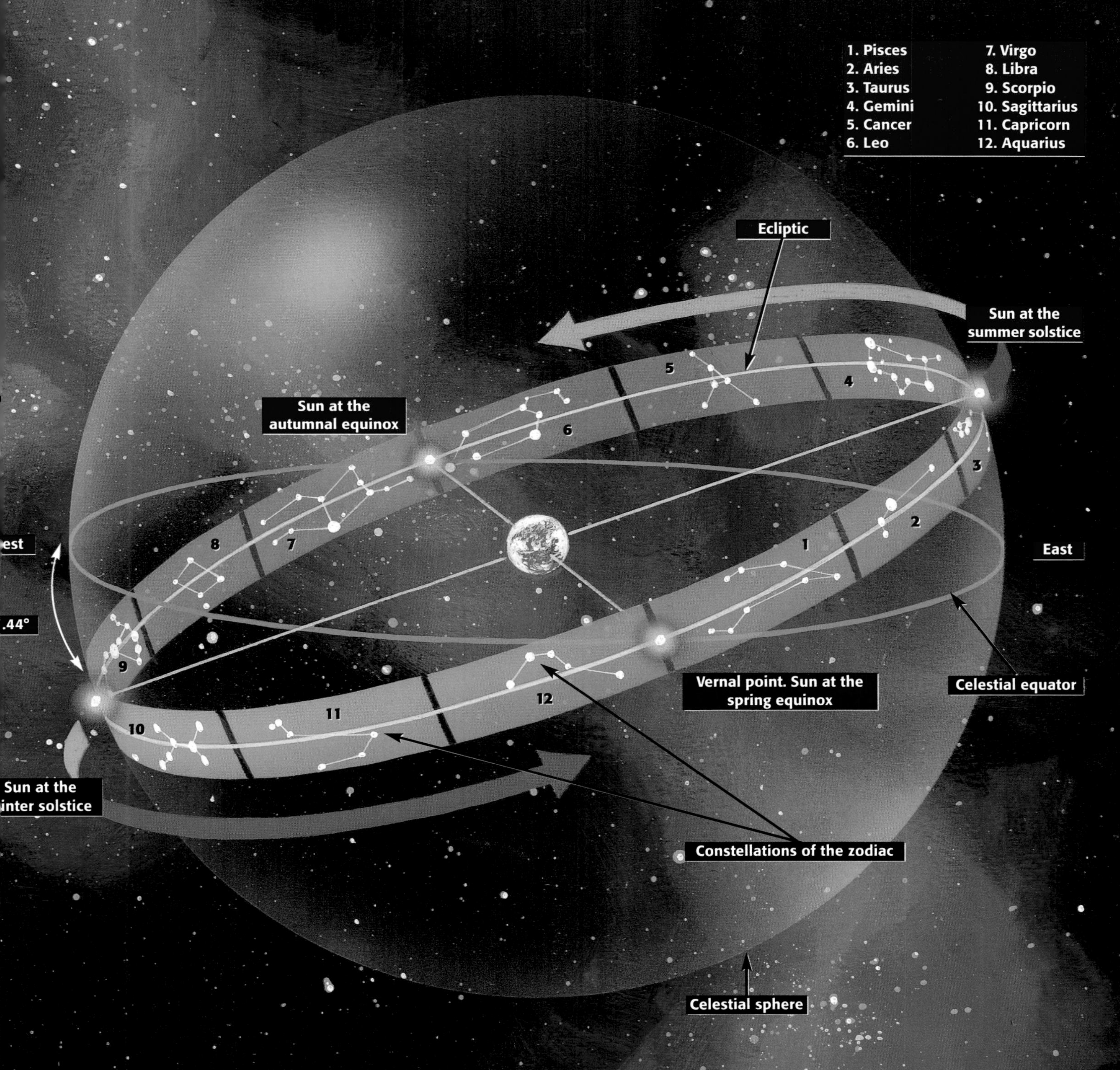

1. Pisces
2. Aries
3. Taurus
4. Gemini
5. Cancer
6. Leo
7. Virgo
8. Libra
9. Scorpio
10. Sagittarius
11. Capricorn
12. Aquarius
Ecliptic
Sun at the summer solstice
Sun at the autumnal equinox
est
East
.44°
Vernal point. Sun at the spring equinox
Celestial equator
Sun at the inter solstice
Constellations of the zodiac
Celestial sphere
1
2
3
4
5
6
7
8
9
10
11
12

sphere. We need only consider that our favorite star is not frozen like its colleagues, but glides along its own interior wall. Our glorious star behaves exactly like a snail crawling along a fence. If the fence pivots, drops or starts to lean dangerously, the gastropod pivots, drops and leans with it. It is obliged to follow the movement of the wall, is dependent upon it, yet can creep along as it pleases. It is the same for the sun. In a year, it traces, across the sphere and in a west-east direction opposite that of our planet's rotation, a circle around the earth. It is tilted in relation to the celestial equator, an extension of that of our planet, at an angle that measures exactly that of the tilt of the terrestrial axis, or 23.44 degrees. In short, the ecliptic is a perfect replica, transferred over to the celestial sphere, of the apparent annual movement of the sun (see sidebar, p. 42). While traveling along this route, the sun, like the snail, is subject to the rotation of the sphere that compels it to rise each morning.

Celestial globes, which reproduce the contents of the sky with its stars and constellations, already existed during Hipparchus's lifetime. Compared to a terrestrial map, the figures in the sky are reversed: they are depicted as they would be seen from outside the sphere, rather than from the earth.

[Vincenzo Coronelli, wood globe. Paris, BNF. © BNF]

Obliquity of the Ecliptic

The ecliptic has long been used as a circle of reference to determine the position of a star and the nearby planets (see chapter 6). Thanks to the constellations of the zodiac (see sidebar, p. 42), the first astronomers were able to locate the point on the ecliptic below or above the one on which a celestial body was found: two fingers below the third star of Leo's tail, or one hand above the eye of Scorpio. But the obliquity of the circle of the apparent solar orbit, that is, its tilt in relation to the earth's equator, varies. It measures today 23.44 degrees; in the second century B.C., the Greek astronomer Hipparchus gave it a value of 23.83 degrees. Ten centuries later, the Arab al-Battani found an obliquity of 23.48 degrees, or .08 degrees more than that measured around 1590 by Tycho Brahe. In 1750, English astronomer James Bradley settled upon 23.46 degrees. They all were right. In fact, the obliquity of the ecliptic varies over time, and currently shrinks by .013 degrees per century.

THE EQUATORIAL COORDINATES

This shift is fairly weak, so the ecliptic, marked out by the zodiacal constellations, remains an excellent celestial reference point for those who wish to keep track of celestial bodies (see chapter 6). More interesting to astronomers is a very particular point on this fictitious solar trajectory: the vernal point, the spot on the ecliptic where the daystar is found during the spring equinox (see sidebar, p. 42). This point, to which the sun returns each year, occupies a fixed location on the sphere, just like the Greenwich observatory firmly planted in the earth's soil. The heavenly ball can thus be marked out, like our planet, by a network of celestial meridians stretching from one pole to another. They run perpendicular to the celestial parallels, which lie parallel to the celestial equator. The celestial meridian, which serves

as a basic point of reference, passes through the two celestial poles and the vernal point. The position of Antares in relation to this point and to the sky's equator does not vary a hair's breadth, and they are carried along together with the movement of the sphere. This line and the circle of the equator, which comprise the equatorial coordinate system (see sidebar, below), have allowed astronomers to assign unique and definitive residences to the stars, delivered key-in-hand in ephemerides, celestial atlases and sky maps.

THE EQUATORIAL COORDINATES

The system of equatorial coordinates is a faithful replica of the one used on earth. The sphere of fixed stars is cut up into circular slices parallel to its equator, the parallels. The celestial equivalent of the meridians runs perpendicular to them and passes through the two poles of the sphere. To assign a precise position to a star, you must find the meridian and parallel at whose intersection the star is located. The meridian will be defined by the right ascension, which resembles, roughly, terrestrial longitude: it is the angle, measured at the equator, that the celestial meridian makes in passing through the star with a meridian taken arbitrarily as a source, like the geographic meridian in Greenwich. This circle of reference is the one that passes through the celestial poles and the vernal point, the intersection of the ecliptic with the celestial equator where the sun is located during the spring equinox. The right ascension is measured from west to east starting from this point. It is not given in degrees, but in hours, minutes and seconds. The sun, dragged along by the sphere, completes a full circle around the earth in twenty-four hours. A circle consists of 360 degrees; ancient astronomers used time and angles as equivalents. Astronomers today have kept this notation: the hour is an angle valued at 15 degrees. As for the parallel, it is determined by its declination, that is, by the angle made by a straight line connecting the star to the center of the sphere of fixed stars—the earth—with the plane of the celestial equator.

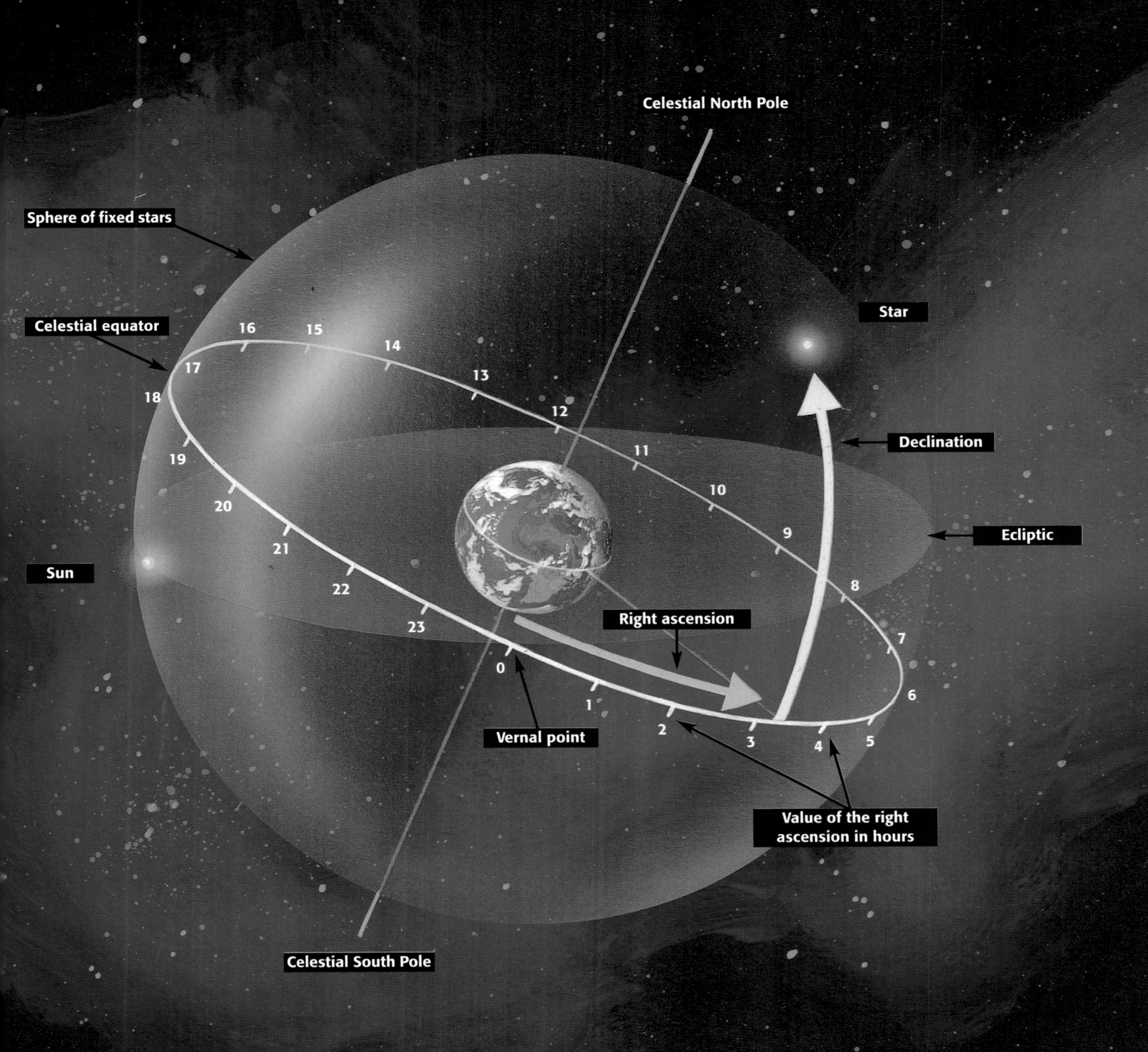

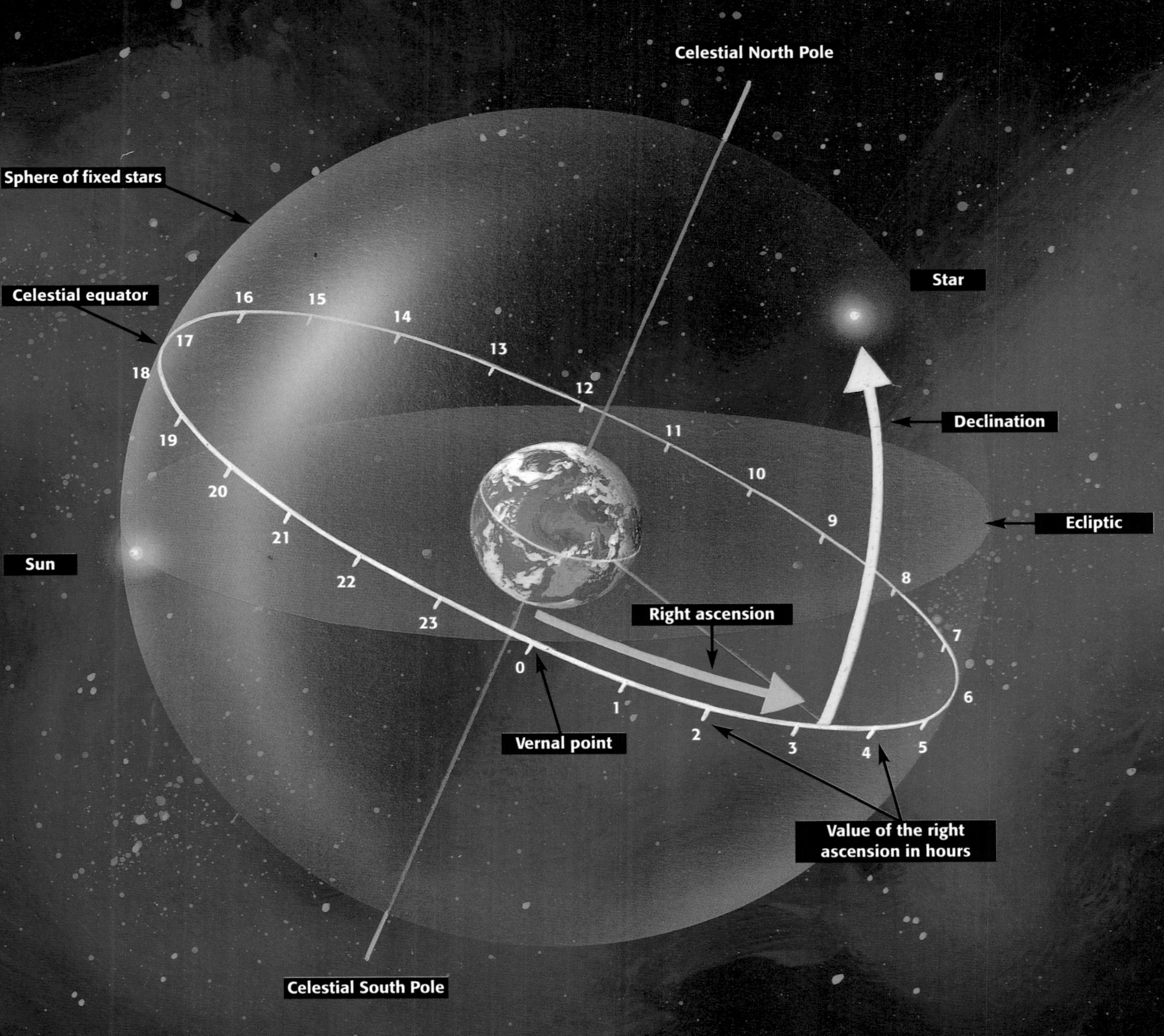
Celestial North Pole
Sphere of fixed stars
Celestial equator
Star
Declination
Ecliptic
Sun
Right ascension
Vernal point
Value of the right ascension in hours
Celestial South Pole
0
1
2
3
4
5
6
7
8
9
10
11
12
13
14
15
16
17
18
19
20
21
22
23

3 Mapping the Universe

1. THE GREAT ANCESTORS 2. UNDER THE SUN OF ISLAM 3. TYCHO BRAHE, PRINCE OF STARS 4. ARTIFICIAL EYES

Hardly had the world been wrapped up in a neat little ball when a handful of Greeks rolled up their tunics and started off on the hunt for stars. Armed with only their eyes and a few rudimentary instruments for measuring angles, they began taking down one by one the positions of the several thousand stars visible to them. Their knowledge then spread to the Muslim world, which continued the meticulous inventory of the celestial sphere's population in medieval observatories that flourished from Cairo to Samarkand.

Passion for the stars reached its zenith in sixteenth-century Uraniborg, Denmark, the fiefdom of Tycho Brahe, who unquestionably ranks among the greatest observers the world has ever known. In the 1600s, thanks to the invention of the telescope, myriads of new stars leaped forth from the darkness to take their place beside the ancient known ones. Finely worked celestial globes and star atlases that, with the

Early astronomical instruments were very rudimentary. Jacob's Staff, a simple cross made up of two small rulers, was used to estimate angles between objects. The smaller rod slid along the larger one until its length roughly corresponded to the distance between two celestial bodies.

[Germany, 1530. © AKG, Paris]

passing centuries, had become veritable works of art, could no longer contain this world. Eventually, the stars—reduced to mere black spots on photographic film—got lost within the thousands of pages that make up modern astronomic catalogues, without even the outline of a constellation to support them . . .

1.

THE GREAT ANCESTORS

Until the sixteenth century, Hipparchus and Ptolemy alone had created practically all of Western astronomy. The former was an astute observer and author of the first-known large-sized stellar catalogue. Given his meager resources, the importance of his astronomical discoveries is truly astounding. The latter is famous most of all for the odd trajectories he forced celestial bodies to follow around the earth (see chapter 6). His catalogue of stars remained for centuries the principal reference guide for observers.

HIPPARCHUS

Born in Nicaea, Hipparchus lived and observed on the island of Rhodes between about 160 and 127 B.C. If you believe the Roman historian Pliny, a supernova drove Hipparchus to conduct a census of the celestial sphere's population. A supernova is a star that has exploded, releasing a phenomenal amount of energy. It becomes as bright as a comet for a certain time, and then is extinguished forever. In 135 B.C., Hipparchus realized that a new star had just appeared in a spot where he had previously seen only darkness. He began to wonder if this phenomenon was really new and if the stars were even fixed at all. Pliny tells us: "He dared do something that was rash even for a god: that is, count the number of stars for the benefit of his successors and revise by name the list of constellations."

Classification of the Stars

In the catalogue he completed around 129 B.C., Hipparchus carefully recorded the position of 850 stars across the celestial sphere. In keeping with the ancient Greek love of order and harmony, he classified the stars according to a system of his own invention: their magnitudes. He managed to define six broad categories of stars based only upon variations in their discernible brightness. The brightest stars are said to be of first magnitude, those slightly less bright of second magnitude, and so on in order of decreasing luster down to the faint stars of sixth magnitude, or those barely perceptible to the naked eye.

The Precession of the Equinoxes

In compiling the measurements carried out by his Greek and Mesopotamian predecessors, Hipparchus came to realize that the celestial sphere wobbles on its axis. He compared his own calculation of the position of the star Spica, in the constellation Virgo, with the one made by a certain Timocharis between 294 and 283 B.C. Both relied on the ecliptic to pinpoint the positions of stars. The ecliptic played the same role as the celestial equator in the system of equatorial coordinates (see sidebar, p. 46): the longitude of a star corresponds to the angle, measured at the ecliptic, that separates its celestial meridian (the line passing through the two poles of the

A Spanish version of Ptolemy's catalogue. It lists the stars constellation by constellation (here, those of Taurus) and provides their coordinates and magnitudes.

[Giovanni Paolo Gallucci, *Theatro del mundo di el tiempo*, 1612. Paris, BNF. © BNF]

celestial sphere and the star) from the meridian at the vernal point. The one Hipparchus determined for Spica differed two degrees from that measured by Timocharis. His ingenious conclusion: the starting point for his measurements, the vernal point, had moved back two degrees in relation to the position it occupied 160 years earlier. The plane of the celestial equator and that of the ecliptic no longer intersected at the same spot as in Timocharis's era, but a little bit further up. Hipparchus interpreted this phenomenon, called precession of the equinoxes, as the result of a slow movement of rotation of the celestial sphere's axis around a point vertical to the plane of the ecliptic (see sidebar, pp. 52–53). In reality, it's the earth's axis that shifts, making the plane of the equator wobble in relation to the ecliptic. Because of this movement, the stars' equatorial coordinates have to be revised regularly. To correct them, you need only know the date on which the position of the vernal point used for reference was calculated.

PTOLEMY

Hipparchus's work is known to us mostly through later authors. Foremost among these was the mathematician, geographer and astrologer Claudius Ptolemy, who made frequent references to Hipparchus in his own writings. This other superstar of astronomy, who lived around A.D. 100 to 170, resided in Alexandria, a vibrant Greek intellectual haven under Roman rule at the time. Ptolemy's thirteen-volume *Mathematical Compilation*, more commonly known as the *Almagest*, is an encyclopedia of practical and astronomical knowledge of the era, which the author reviewed and corrected in light of his own speculations, calculations and observations. His description of the movement of celestial bodies, the

THEATRO DEL MVNDO

TAVRO CONSTELLACION 23. Cap. 27.

	Forma y nombres de estrellas		Longitud G	M	Latitud G	M	Magni.	Na
1	En la Section, o diuission de las quatro la boreal	A	19	40	6	0	4	♀
2	la otra despues desta.	A	19	20	7	15	4	Y
3	la tercera dellas,	A	18	0	8	30	4	po
4	la quarta q̃ cae al Austro	A	17	50	9	15	4	co
5	En la espalda derecha,	A	23	0	9	30	5	
6	en el pecho,	A	27	0	8	0	3	♄
7	en la rodilla, derecha	A	30	0	12	40	4	
8	en la quartilla derecha,	A	26	20	14	50	4	
9	En la rodilla izquierda,	A	35	30	10	0	4	
10	en la quartilla izquierda,	A	36	20	15	30	4	
11	de las cinco que estan en el rostro llamadas Suculas, o Hyadas la de la nariz	A	32	0	5	40	3	
12	la q̃ esta ẽtre esta y el ojo boreal	A	33	40	4	5	3	
13	Entre la mesma y el ojo austral	A	34	10	5	50	3	♂
14	la luciẽte en el mesmo ojo vajo la ceja q̃ llamã oculus Tauri	A	36	0	5	10	1	
15	la del ojo boreal	A	35	10	3	0	3	
16	la del nacimiento del cuerno, y oydo Austral,	A	40	20	4	0	4	
17	La austral de las dos en el mesmo Cuerno	A	43	4	5	0	4	Y
18	la que mas al boreal	A	43	20	3	30	5	po
19	ẽ lo extremo d̃l mesmo cuerno	A	50	30	2	40	3	co ☿
20	en el nacimiento del Cuerno Septentrional,	B	40	0	4	0	4	
21	En el extremo del mesmo cuerno q̃ cae en el pie d̃recho de Ericthonio,	B	40	0	5	0	3	♉
22	de las dos d̃ la oreja la boreal	B	35	20	4	30	5	♄
23	la austral dellas	B	35	0	4	0	5	
24	la precedẽte de las dos peq̃ñas de la Cerbiz,	B	30	20	0	40	5	Y po
25	La siguiente	B	32	20	1	40	6	co
26	en el cuello de las.4. precedentes del lado la austral,	B	31	20	5	40	5	☿
27	la boreal del mesmo lado	B	32	0	7	20	5	
28	la austral del siguiente lado,	B	35	20	3	[illegible]	5	
29	La boreal del mesmo lado	B	35	0	5	0	5	☉
30	el termino boreal del precedẽte lado de las Pleyades.	B	[illegible]3	30	[illegible]	30	5	po ♃

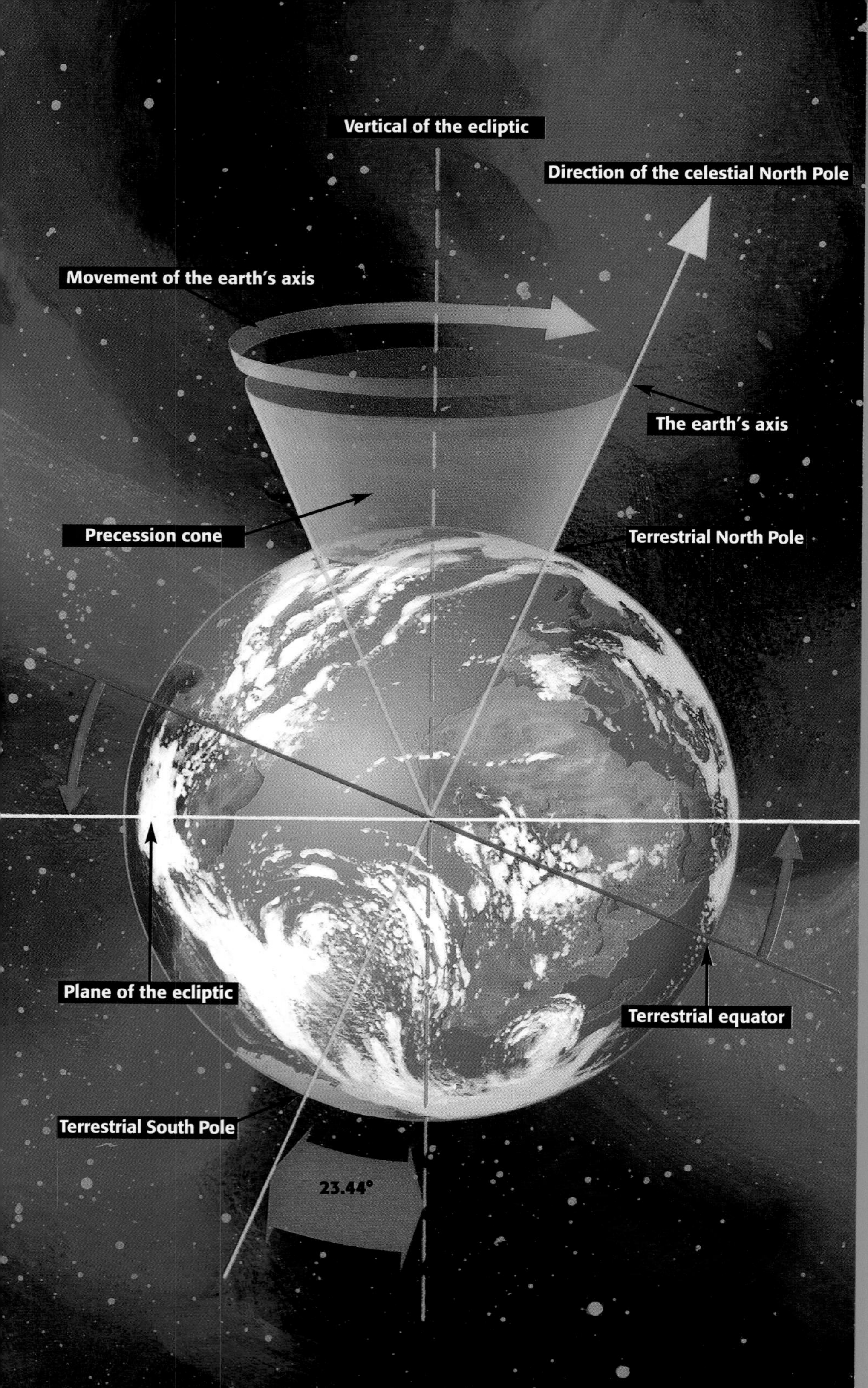

THE PRECESSIO

The earth's axis doesn't always occupy the same position in relation to the stars. Today, it points toward the North Star, in the tail of Ursa Minor (the "Little Bear," a.k.a. the Little Dipper), which marks the northern celestial pole. Five thousand years ago, this position was held by the star Thuban in the constellation Draco. In twelv
thousand years, this role will be played by th
brilliant star Vega in the constellation Ly

It's not the stars that are shifting but t
earth's axis. It's like the stem of a top as
winds down its whirl: it spins, makes a circle
the sky and doesn't return to its initial pos
tion, aiming at a given star only once ever
twenty-six thousand years. The point at the center of the circle drawn by the earth's axis is the ecliptic North Pole: that is, the point of intersection between the vertical of the plane of the ecliptic and the celestial sphere. The earth's axis is perpendicular to the plane of the equator. It roughly resembles a

F THE EQUINOXES

toothpick stuck into a cucumber slice: if we have it spin around a different axis, the slice is automatically carried along in the movement. The plane of the earth's equator shifts in relation to that of the ecliptic.

Thanks to the corkscrew principle, which allows these movements to be con- idered from a reverse perspective, we can nagine the axis of the sphere of fixed stars the direct extension of that of the spinning rth. It causes the celestial equator to wob- e and sway. The vernal point, situated at he intersection of this plane with that of the ecliptic, shifts due to this wobble: it migrates—astronomers say that it retrogrades—very slowly westward. It makes a complete trip around the earth's equator in twenty-six thousand years. In Hipparchus's era, on the spring equinox the sun entered the zodiacal constellation of Aries. Today, the vernal point is in the constellation Pisces.

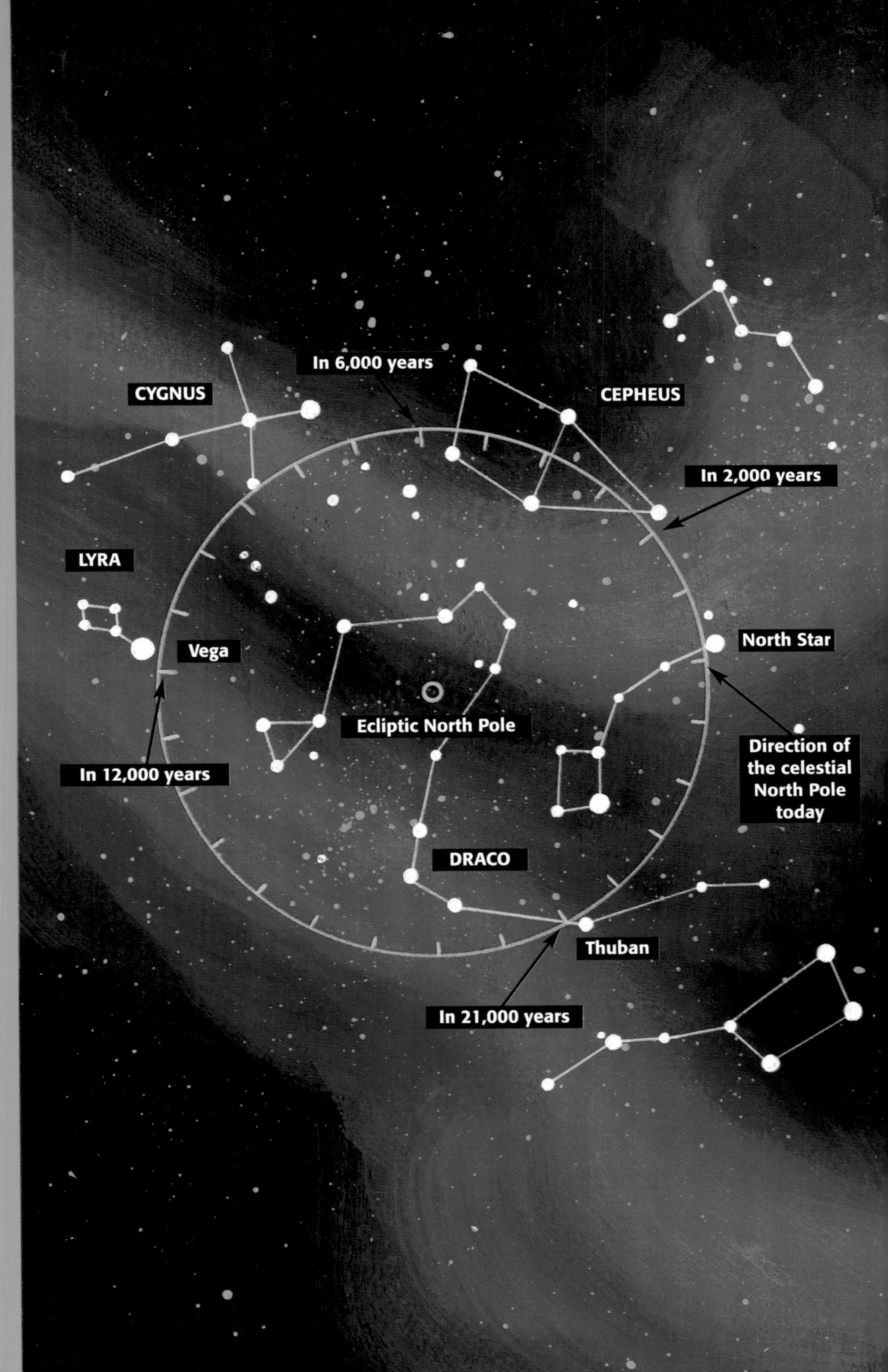

weird circular trajectories he forced the sun, moon and planets to follow around the earth, are the master strokes of the Almagest (see chapter 6). Though incorrect, this model managed to escape the wrecking ball for over a thousand years, thanks to the impressive precision it afforded calculations and predictions. It can't be denied that Ptolemy proceeded in an orderly, methodical fashion, describing in detail both the mathematical and practical tools—the instruments—he deemed useful for understanding the universe, and insisting upon their proper use and the necessity of constantly comparing calculations to observations.

He himself relied on measurements recorded by the Mesopotamians to refine his models, and adopted those of Hipparchus for a stellar catalogue in which he gave the magnitudes and positions, in relation to the ecliptic, of 1,028 stars divided among forty-seven constellations.

The astronomer Abd al-Rahman al-Sufi (903–986) is the author of a famous treatise, *The Book of Constellations and Fixed Stars,* in which he illustrated and described in minute detail the forty-eight constellations inventoried by Ptolemy. Translated into Latin in the twelfth century, his work did much to establish and spread the use of Arab names for the stars.

[Abd al-Rahman al-Sufi, *The Constellation Centaurus Fighting Leo,* from *The Book of Constellations and Fixed Stars.* Paris, BNF. © BNF]

2.

UNDER THE SUN OF ISLAM

Upon the death of the prophet Mohammed in A.D. 632, the knights of Islam left Arabia to complete a sacred mission: to deliver the message of the Koran to the four corners of the earth. They very quickly found themselves at the head of an enormous empire, which stretched from Spain to the gates of India. In 762, Caliph al-Mansur, brother and successor to Abu al-Abbas, founder of the illustrious Abbasids dynasty, decreed beautiful Baghdad its capital. His grandson, the mythic Caliph Harun al-Rachid, began collecting ancient manuscripts. Their translation began in 813, with the reign of his son, Caliph al-Ma'mun, who equipped Baghdad with a great cultural and scientific center called Bayt al Hikma, or "House of Wisdom," and assembled a team of leading scientists—bringing together Muslims, Christians and pagans—who provided him with four Arabic versions of Ptolemy's *Mathematical Compilation*. It was renamed *Al-majusti*, a Greco-Arabic concoction meaning "the Great One" and that became *Almagest* in Latin.

THE STARS RENAMED

Al-Ma'mun's studious enterprise marked the sky with the indelible seal of the Arabic language: the brilliant stars kept the names given them by the tenth-century astronomer Abd al-Rahman al-Sufi, the translator of Ptolemy's stellar catalogue. Betelgeuse, for example, is a distortion of *Ibt al Jauza*, which means "Giant's Armpit." Algol, in the constellation Perseus, is *Ras al-ghul* or "Monster's Head": it represents the eye of Medusa, the hideous creature decapitated by the hero Perseus. These names were transmitted to the West—which has retained them—thanks to an instrument as beautiful and mysterious as a jewel: the astrolabe. Developed by the Arabs according to Ptolemy's posthumous directions, which perhaps he himself had borrowed from Hipparchus, the astrolabe provides a two-dimensional representation of the celestial vault as it appears at certain latitudes. It was used as both a watch and a compass.

The underground section of the enormous sextant at the observatory in Samarkand was rediscovered by accident in 1908. An arm slid along rails lined with sheets of marble and measured the height of stars in the sky.
[© The Bridgeman Art Library]

Depending on the mood and erudition of the loca potentates, numerous observatories flourished through-out Egypt, Turkey and Iran. Some were highly organized such as the one in Maragha in Iran, founded in 1259 by the Mongol Khan Hulagu: astronomers there worked in teams, shared tasks and pursued a rigorous program o study and observation. The majority of these establish-ments, however, had very short lifespans. One of the most ephemeral was built in Cairo in 1120. The vizier in charge was executed five years later for entering into communication with Saturn (astrology is looked upon very poorly by orthodox Muslims), and the observatory disappeared just as quickly.

Ulug Beg, Astronomer King

The last great observatory of the Muslim world was the work of an astronomer prince, Muhammed Turgay more commonly known by the name of Ulug Beg, or "Great Prince." This grandson of Tamerlane was at age fifteen named governor of Transoxiana, a territory cor-responding roughly to current Uzbekistan and whose capital was none other than Samarkand. Around 1420 he had an observatory built on a hill north of the city It was shaped like an enormous cylinder, 115 feet high and 157 feet wide. It housed the largest instrument ever seen at the time, the Fakhri sextant: a 60-degree arc made of stone and measuring 130 feet in diameter. One part of the instrument sank 36 feet underground Aligned with the north-south meridian of Samarkand it was used to measure the height of stars in the sky. The *Zij I Gurgani*, Ulug Beg's *Astronomical Tables* published in 1437, provided the positions of 1,018 stars with such precision that it represented the bes catalogue of its era. The astronomer prince was decapitated on 17 October 1449 with the tacit approval of his own son.

3.

TYCHO BRAHE, PRINCE OF STARS

Europe remained unaware of the existence of Ulug Beg's *Tables* until 1650. Western astronomy would have been saved a lot of time had they been known, for it took the West a century and a half to acquire an observatory and measurements to rival those of Samarkand. It all happened thanks to Tycho Brahe. Born in 1546, Brahe was a product of the Danish aristocracy. As a little boy, he was kidnapped by his uncle, Vice Admiral Joergen, who took charge of his education and made him his heir. Tycho was still a teenager when his adoptive father died of pneumonia, contracted after jumping into the water to save King Ferdinand II from drowning.

Tycho Brahe was without a doubt the greatest naked-eye observer in history. It was the extraordinary precision of his measurements of Mars's movements that allowed astronomer Johannes Kepler to uncover the true face of the solar system.

[Nationalhistoriske Museum, Frederiksborg Slot. © AKG, Paris]

THE NEW STAR

Legend has it that Tycho fell in love with the stars after observing a partial solar eclipse. In 1572, a new star, even more brilliant than heavenly Venus, appeared like magic in the sky. This was a supernova, which Tycho Brahe carefully observed in all its glory before it disappeared. A year later, he published *De nova stella*, in which meticulous observations of the new star and hypotheses about its actual distance earned Brahe a certain renown.

Uraniborg, the splendid castle Tycho Brahe had constructed on the island of Hveen, had it all. Besides its observatories, which were littered with instruments of his own design, it featured an alchemy laboratory, a paper mill and a printing house, various studios and a windmill. Tycho reigned as a perfect despot over this little world, terrorizing his assistants and brutalizing his farmers, who eventually got their revenge by completely destroying Uraniborg.

[Joan Blaeu, *Atlas major, sive cosmogrphia Blaviana*, Amsterdam, 1662. Paris, BNF. © BNF]

URANIBORG

In 1576, Frederick II of Denmark, the son of Ferdinand who had been saved from the waters by his uncle Joergen, repaid his father's debt in a sumptuous fashion. He offered Tycho the isle of Hveen, an area of 760 hectares located between Copenhagen and the castle of Elsinore, along with the inhabitants residing there. The astronomer had Uraniborg built on the island: the prideful dream of an eccentric, astronomy-mad megalomaniac, its detachable roof spiked with its towers and its interior packed with instruments.

Tycho endeavored to entirely overhaul existing sky maps and engraved, one by one, on an enormous celestial globe the positions of 777 stars, measured with a precision unrivaled at the time. He also observed the planets at length; his measurements of the positions of Mars are considered the best of his era.

THE MAN WITH THE GOLDEN NOSE

As a student, Tycho Brahe was beaten in a duel by a schoolmate who claimed to be better than him in mathematics. A sword blow removed a piece of his nose, which he replaced with a prosthesis made from a gold and silver alloy. For more than twenty years, the lord of Uraniborg was the libidinous nightmare of servant girls and gargantuan host to a ceaseless parade of princes, courtiers, kings and scientists beneath the castle's pompous, astronomy-themed frescos. A genuine despot, Brahe roared at the sinister jokes of his fool, the dwarf Jepp, while crushing his serfs with taxes and having their families locked in irons. The young King Christian, Frederick II's heir, was eventually moved by the terrible plight of Hveen's peasants and decided to reduce Tycho's privileges. The astronomer preferred to abandon his island than submit to such privations. Leaving Uraniborg to the hands of the farmers, he moved his court and all his instruments into the castle of Benatek, several miles from Prague. He died there on 24 October 1601, possibly from a rupture of the bladder, eleven days after attending a princely supper.

4.

ARTIFICIAL EYES

In 1609, Italian physicist and astronomer Galileo Galilei had the clever idea of pointing toward the sky a device invented several years earlier in the Netherlands: the telescope. There he saw "below stars of sixth magnitude, a mass of other stars that escape natural vision, so numerous that it can hardly be believed, for one can observe more than six additional categories of magnitude."

A MAD WORLD!

Celestial maps and atlases had already been enriched in the fifteenth and sixteenth centuries with armfuls of bizarre constellations—the air pump, the telescope, the compass—reported by navigators returning from their incursions into the lands and seas of the Southern Hemisphere. In revealing thousands of stars that had previously remained invisible, the telescope also multiplied the number of new constellations. Polish astronomer Johannes Hevelius, for example, cheerfully offered the sky a lynx, a little fox (in whose mouth he placed a goose, no longer there today), a lion cub, hunting dogs and, in gratitude to his patron, monarch

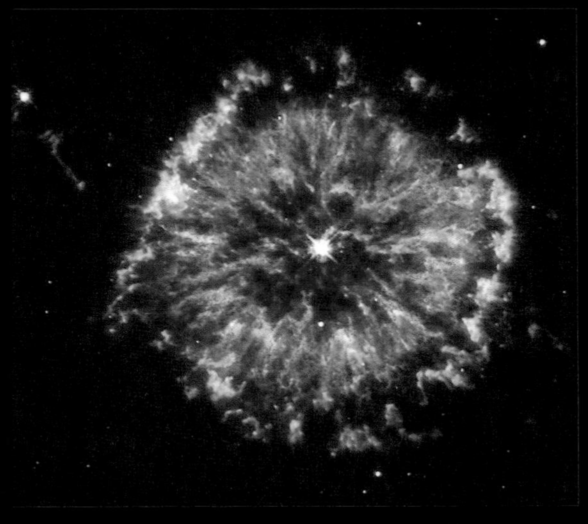

The planetary nebula NGC 6751. When astronomers discovered that there are more than just stars and planets in the sky, they called all new bodies "nebulae," because of their somewhat fuzzy and cottony contours. Today, the term designates specifically clouds of gas found in interstellar space or which surround stars. The latter clouds are called planetary nebulae.

[© *Ciel & Espace*/NASA/Hubble Heritage]

Johannes Hevelius's azimuthal dial (1673). It was used to determine the horizontal coordinates of the stars.
[© AKG, Paris].

John III Sobieski, placed a curious Sobieski's Shield in the firmament. He wasn't always so inspired, however, and in 1690 transformed the vaporous Veil of Veronica, hung out to dry by a colleague between the Leo and the Hydra, into the rather prosaic Sextant. The International Astronomical Union wouldn't bother tidying up the sky until 1922. They held onto eighty-eight constellations—forty-seven of which came directly from Ptolemy—whose boundaries Belgian astronomer Eugène Delporte set definitively and scientifically in 1930.

NEBULAE

Astronomy, until then practiced by a few isolated individuals subject to the capricious erudition of rich princes and patrons, became an affair of state and a veritable profession upon the foundation of the first great public observatories: the one in Paris opened in 1667, followed by Greenwich in 1675, Berlin in 1705 and Moscow in 1750. The first telescopes began poking out their noses at the end of the 1600s. In the following century, Charles Messier, nicknamed the "Ferret of Comets" by Louis XV, set off hunting these beautiful, long-haired "stars" and returned with nothing better in his nets than cottony clouds, prudently named *nebulae*. In 1784, he published a list containing 103 of these new celestial bodies, at the time labeled "Messier's Objects." At the start of the following century, William Herschel, discoverer of the planet Uranus and maniacal perfectionist whom his sister Caroline had to spoon-feed while he polished the mirrors of his telescopes, counted 2,500 of them. Nebulae appeared sometimes as galaxies, sometimes as mere gaseous shells spit out by stars, or as clouds of gas, or even breeding grounds for planets.

HIPPARCOS

These newcomers quite naturally took their place in the celestial atlases that, now stripped of the chimeras of mythology, jumbled the stars up by the tens of thousands. Photography burst onto the scene at the end of the nineteenth century and increased their number even more. Admiral Mouchez, director of the Paris Observatory, launched a fantastic project in 1884 designed to create a photographic map of the sky and that eventually provided the positions of two million stars!

The catalogues listing all the stars' characteristics became thick and heavy as telephone books. With photography, and then the automation and computerization of instruments of observation, astronomers kicked the habit of observing the sky directly and entrusted the meticulous counting of its population to the machine. HIPPARCOS (acronym of the High Precision Parallax Collecting Satellite) is a satellite put in orbit around the earth on 8 August 1989. It has reviewed over a million celestial objects and provides their positions using the system of galactic coordinates: it's now the plane of the galaxy's equator that serves as its reference point, and no longer that of the celestial sphere. Its precision is far beyond that of its illustrious and very human namesake.

The telescope on Mount Palomar, in California, started operating in 1949. Capable of detecting a candle over 15,000 miles away, it remained the largest and most powerful telescope in the world for almost thirty years.

[Russell Porter, drawing, 1939. © California Institute of Technology, Pasadena]

ASTRONOMY'S ANGLE

Apparent diameter

A circle measures 360 degrees, with 1 degree divided into 60 arc minutes (noted ') and 1 arc minute into 60 arc seconds (noted "). Astronomers swear by angles alone, which they use to estimate the distances separating celestial bodies and their size. The further away an object is from us, the smaller it appears: it takes only a finger to obscure an oak found a hundred yards away. Seen from a greater distance, the same tree will shrink to only a half a finger in height. If we move closer to it, the tree will reach the size of an arm. It's the same for the moon: a disc one-fifth of an inch in diameter held 22.5 inches from the eye is all it takes to mask her silvery face. At around 32 feet, you'll need a token of about 3.5 inches across. The further you move the disc away from your eye, the larger it needs to be to obscure the moon. Rather than amuse themselves estimating a celestial body's size by comparing it against a multitude of coins placed at a multitude of distances from their eyes, astronomers have developed a far more effective trick. They imagine the disc of the moon as the base of a triangle whose point is located at the observer's level, and whose sides equal the earth-moon distance. The angle at the point of this triangle does not change if we adjust the value of its base and sides proportionately. In other words, if the base measures 2,175 miles and the sides 248,548 miles, or respectively 0.2 and 22.5 inches, our angle remains unchanged. There exists an infinity of values of bases and sides for which the angle is the same. This method is reliable for any star: the angle at the summit of the triangle it forms with the observer is what is labeled the apparent diameter. In the moon's case, this averages 30 degrees. For planets and stars, it's considerably smaller. Through a strange quirk of nature, the moon and sun have the same apparent diameter. In fact, the nocturnal goddess, who is four hundred times smaller than our daystar, is also four hundred times closer. It's this happy coincidence that makes eclipses possible. In the same way, angles serve to evaluate the distance between two stars. The arc linking the two objects becomes the base of a triangle whose point again converges at the observer. The angle formed by the two straight lines separating the observer's eye from each of the two stars represents the angular distance. If you know the diameter of the celestial sphere onto which the stars are pinned, then it's possible, taking into account the apparent diameter of celestial bodies and their angular distance, to calculate their size and the distances between them in miles.

Angular distance

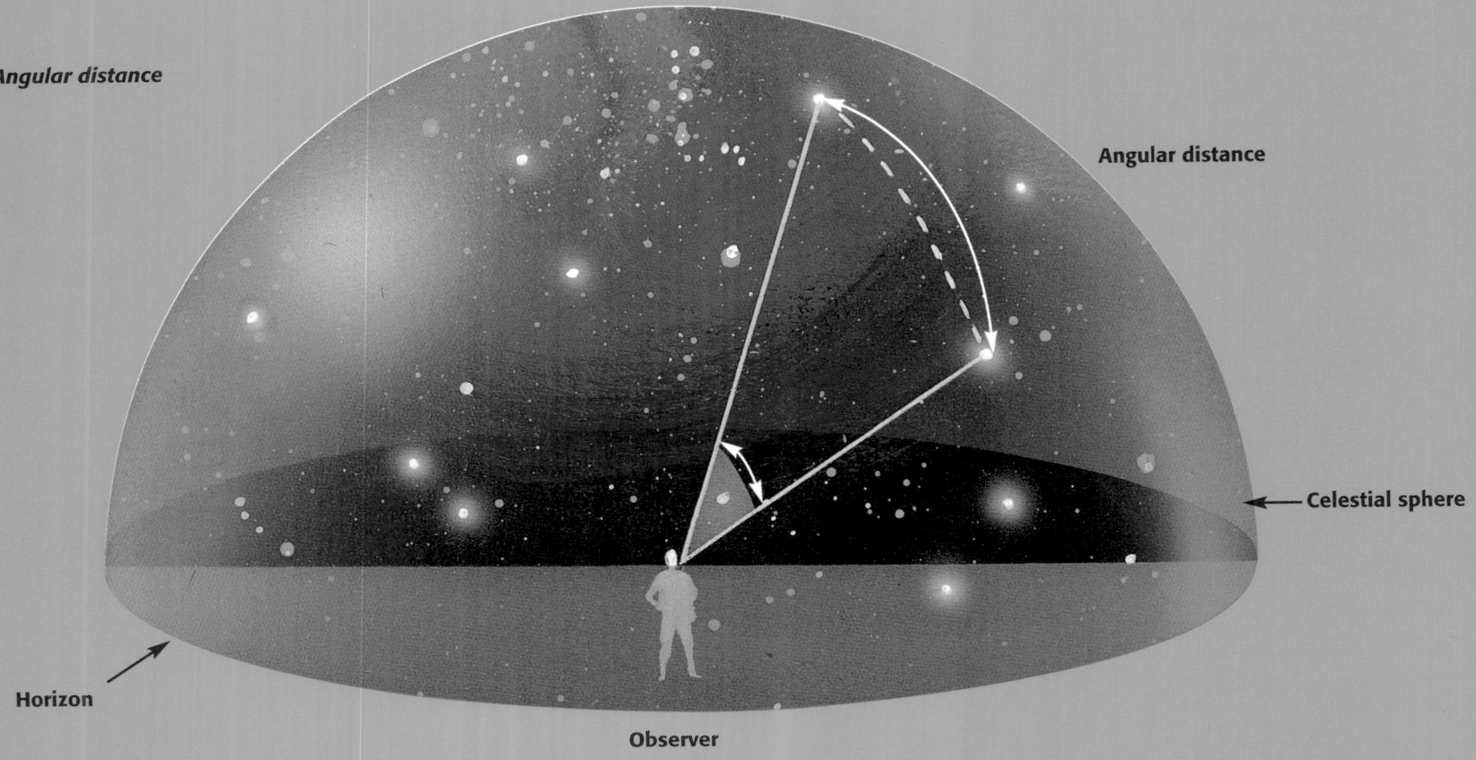

4

Time

1. AT THE TONE, THE TIME WILL BE . . . 2. THE SECRET OF LONGITUDE 3. FROM ONE TIME TO ANOTHER

In the second millennium B.C., long before the legendary Hipparchus pilfered some of their best finds (such as angles), the Mesopotamians strung their sleepless nights together like pearls. They covered thousands of clay tablets with their spidery cuneiform, day after day taking down absolutely every cosmic detail: the risings and settings of the moon, sun, planets and constellations, the look and color of the stars, the positions they occupy in relation to each other.

The Mesopotamians' interest in the heavens was never gratuitous: they were searching for any sign that might reveal the dark designs of the gods. Only in such fertile and convoluted minds could the far-fetched idea sprout that the stars influence the destiny of individuals—though no one's quite sure why or how. The first known horoscopes, from the fifth century B.C., provided the zodiacal positions held by the sun and certain planets on the birthdays of personages who are

now unknown. The Mesopotamians, and after them the Greeks, Romans, Arabs and early Christians, however, had a second, far more mundane and imperative reason for scrutinizing the firmament in this way: time. They treated the sun and stars like the hands of a gigantic clock, glancing at them regularly to tell the time of day or night. Astronomers thus remained the guardians of time for centuries, and had the honor of defining the second based on the movement of the sun. That is, until a vulgar cesium atom came to replace them in the 1970s.

An illuminated allegory of time . . . The sun is time's great organizer. Its annual journey through the constellations of the zodiac sets the rhythm of the passing of the seasons and the year. Its daily trip across the sky was the basis for dividing the day into hours. Those belonging to day wear white, and those belonging to night wear black, the color of mourning.

[*Les Triomphes du poëthe Messire François Petrarche,* sixteenth century. Paris, BNF. © BNF]

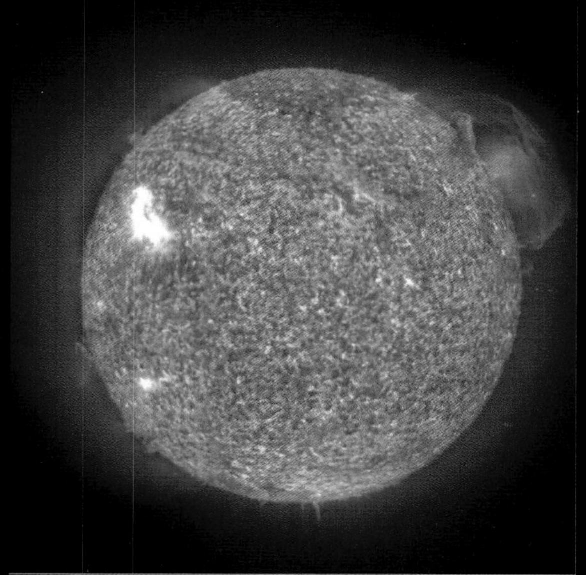

The sun photographed by the satellite SOHO. It's an average-sized star, around five billion years old. This ball of gas in constant fusion, inundating us with its light and heat, is subject to regular magnetic outbursts whose impact on earth we don't yet fully understand.
[© *Ciel & Espace*/NASA/ESA/SOHO/EIT]

1.

AT THE TONE, THE TIME WILL BE . . .

Our Cro-Magnon ancestors possessed two temporal points of reference each day: sunrise and sunset. The former announces the onset of a period of light, day, which comes to a close with the arrival of the latter. The world then enters a dark phase, night, ending with the daystar's reappearance. Between each sunrise and sunset, a varying amount of time elapses.

DIVIDING DAY FROM NIGHT

Endowed with their legendary gift for observation, our forebears likely noticed how the shadow of a stick or totem pole, planted perfectly straight in the middle of their encampment, is fairly long when the sun first appears each day. This shadow becomes shorter as the sun gathers strength in the sky, and then gets longer again toward twilight. They could thus split daytime into two distinct periods: that in which shadows shrink, and that in which they grow. These periods are separated by the brief instant at which the shadow of a stick (a "gnomon" to those in the know) reaches its shortest length of the day.

This system was still current in Greece around 1000 B.C.: Homer, author of the *Iliad* and the *Odyssey,* who was born near Smyrna in the ninth century B.C., acknowledged only the beginning, middle and end of the day. That is, in modern parlance, morning, noon and evening. The Romans went a bit further by recognizing *diluculum* (daybreak), *mane* (morning), *ad meridiem* (toward noon), *meridie* (middle of the day), *de meridie* (afternoon), *suprema* (sunset), *vespera* (evening) and *crepusculum* (twilight).

Bits of Day . . .

The gnomon's shadow creeps along the ground throughout the day. It sweeps across a given stretch of terrain, which can be easily divided into a certain number of sections. One need simply cut the day up into temporal slices: the duration of each slice equals the time taken by the shadow to cover the corresponding fraction of surface. Sundials were invented along this principle. The many little bits of time were eventually parceled out in equal portions and called *horas*—hours—by the Romans.

. . . and Bits of Night

At night, the stars take over for the gnomon. Once the final fires of the sun have gone out, the first constellation appears to the east and rises above the horizon. A second one follows it, followed by a third, a fourth, a fifth. This procession allows specific nocturnal periods to be distinguished: the duration of each period is equal to the time elapsed between the moment a given constellation first appears on the horizon and the moment it completely emerges. This method of breaking down the night with the movement of the stars is very ancient, with certain authors going so far as to claim that the constellations were invented for this purpose alone.

A bronze horizontal sundial from the eighteenth century. Its stylus lies parallel to the axis of the celestial sphere and stands perpendicular to the daily trajectory of the sun. The shadow of the stylus travels across the circle, forever returning to the same position at the same time of day.
[England, eighteenth century. Private collection. © The Bridgeman Art Library]

WHAT TIME IS IT?

The ancients ended up dividing day and night into twelve temporal intervals each, that is, twelve hours. This system was horribly impractical, for each hour's duration varied considerably depending on the time of year.

Unequal hours

In early June, for example, the day lasts a bit more than sixteen modern hours, while in early January, it lasts only around eight. Under the ancient system, both day and night comprised twelve hours each: daylight hours at the beginning of summer lasted twice as long as those in winter. Inversely, summer night hours were almost twice as short as the winter ones. These variations are the result of the sun's movement along the ecliptic. Depending on its position along the celestial sphere, the sun's journey above the horizon—the length of the day—will lengthen or shrink (see side bar, opposite). Such geometrically variable hours are called temporary or unequal, and it was only on the spring and fall equinoxes that night hours were of equal length to those of day. Called equinoctial hours, they were used to calculate the ideal hourly duration.

The clock of the Torre dell'Orlogio in Venice dates back to the end of the fifteenth century. It features only one hand, which takes twenty-four hours—not twelve—to make a full circle around the dial. Its journey across the zodiac mimics the annual travels of the sun.
[© The Bridgeman Art Library]

Three Thousand, Six Hundred Seconds

Day and night are merely two sides of the same astronomical coin: specifically, the apparent and daily "revolution" of the sun around the earth. The term "day" ultimately designates the duration of the full trip, and consists of twelve daylight hours and twelve nighttime hours, that is, twenty-four in total. Why not twenty, thirty or forty? The Romans, from whom we inherited this system, did nothing but copy verbatim the one

THE LENGTH OF THE DAY IN THE NORTHERN HEMISPHERE

At the moment of the spring equinox, around 21 March, the sun is located at the level of the terrestrial equator. For an observer in the Northern Hemisphere, its trajectory above the horizon forms a perfect semicircle. The time taken by the star to make its journey, that is, the duration of the daylight, is exactly the same as that of its nocturnal journey below the horizon. As the days go by, the sun climbs toward the north, and its daily path keeps getting longer: it rises earlier and earlier and sets later and later. On 21 June, the date of the summer solstice, the star is above the celestial equator and is located, in relation to the earth, vertical to the tropic of Cancer. Its journey above the horizon is its longest of the year. In summer, the sun starts its descent southward. It rises later and later and sets earlier and earlier. Around 23 September it crosses the second point of intersection between the ecliptic and the celestial equator. This is the autumnal equinox: day is again equal to night. The former continues getting slimmer while the latter settles in, until the winter solstice arrives. Around 21 December the daystar has reached the lowest point in its trajectory; it shines above the tropic of Capricorn. This is the shortest day of the year in the Northern Hemisphere.

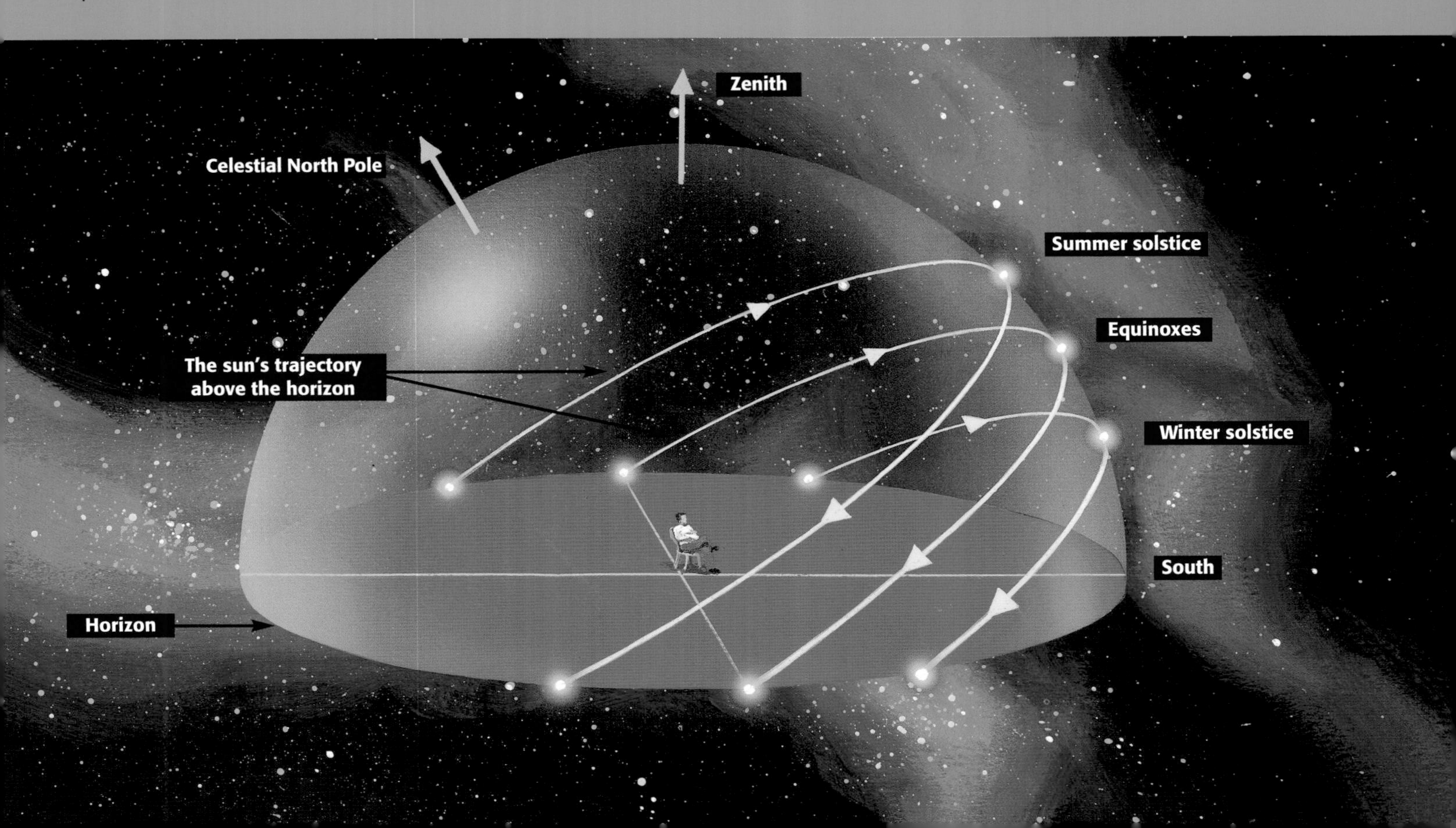

A ninth-century Arabic astrolabe. The backside (not shown here) featured a sight system used to measure the height of the sun or visible stars. The front side, a moveable and finely worked disc—the spider—represents the celestial sphere. Each of its pointers corresponds to a star. By turning the spider, the observer can re-create the appearance of the sky at a given moment and at a precise latitude. From there, the time can be deduced.

[Paris, BNF. © BNF]

already established by the Mesopotamians. The latter worked only according to the sexagesimal system, that is, in multiples of sixty, and divided day and night into six fractions each. This number was later doubled. When astronomers began needing to calculate time with greater precision, hours were divided like angles into sixty minutes, with each minute divided into sixty seconds.

Astronomical Day and Calendar Day

Some civilizations had day begin at sunrise and end the following one. Others, in contrast, preferred to have it begin at sunset. For their part, astronomers set the beginning of day at the precise moment the sun sits directly vertical to the meridian where the observer is standing, or just above the geographic line passing through the planet's two poles and observer's location (see sidebar, opposite). At that moment, it is exactly zero hours in astronomical time. Twenty-four hours later, the sun returns to the same position above the meridian and counters are reset to zero. Moving from one date to the next smack in the middle of the day was not very practical for common folk, so they decided to start the day twelve hours after the astronomers did. Ever since, it is zero hours in civilian time when the sun sits vertical to the antimeridian—the line diametrically opposite that of the meridian—and noon when it passes over the meridian.

Sidereal Time

At the very beginning of the month of February, at 9:00 P.M. Universal Time (UT), observers in the Northern Hemisphere cannot miss one of the most remarkable constellations in the sky: Orion. Imagine you carefully

THE SUN'S PASSAGE OVER THE MERIDIAN

Astronomers say that the sun, or any other star, culminates when it is located at the highest point in its trajectory in relation to a given horizon. In other words, that is when it is sitting right on the imaginary line that passes through the two poles of the celestial sphere and the zenith point vertical to the observer's location. This celestial meridian is the extension, on the sphere of fixed stars, of the observer's geographical meridian, the fictitious line connecting the observer with the two poles of the earth.

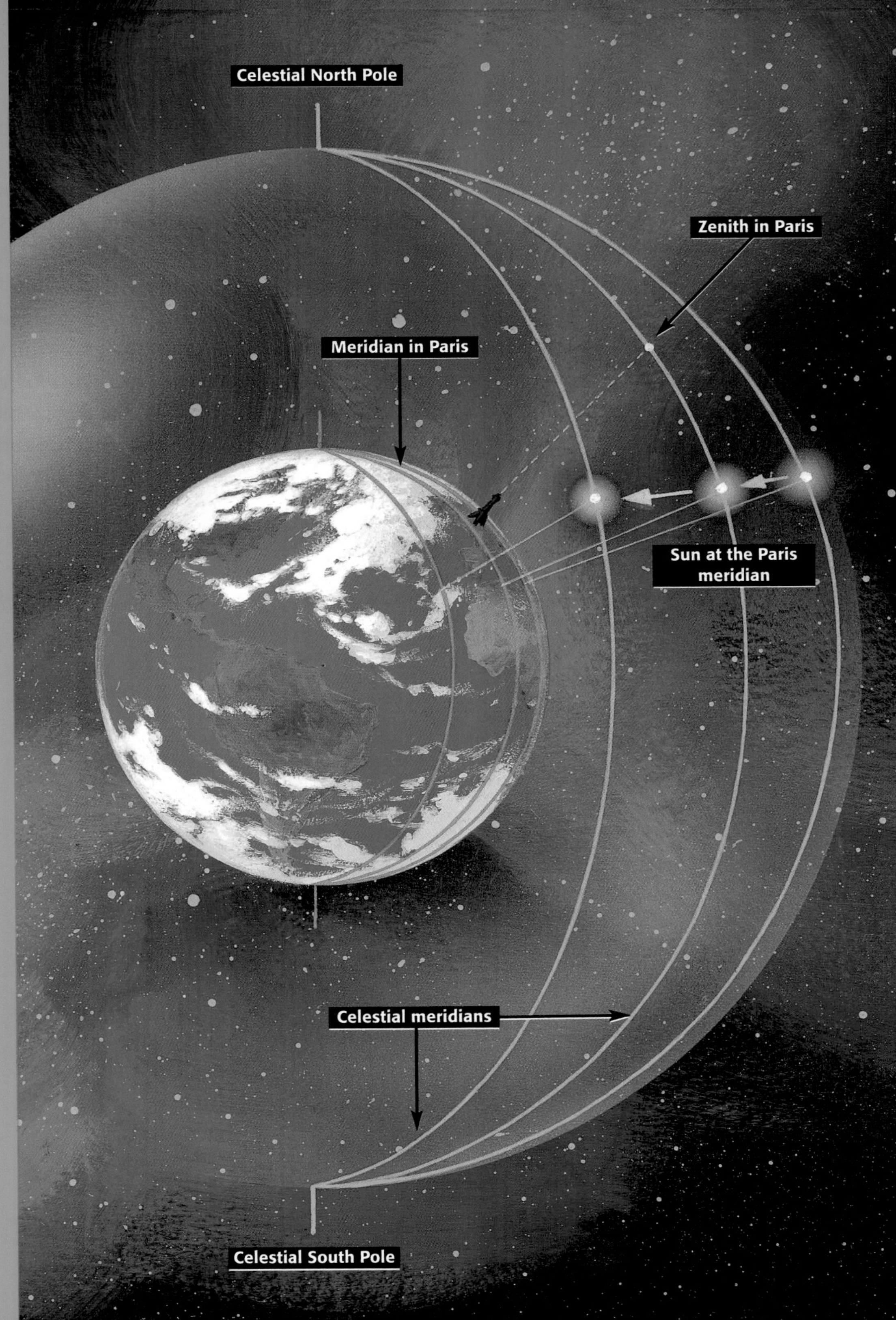

The mechanism of the earliest clocks left much to be desired, as they were never synchronized with the sun. Solar dials were in large part even less reliable. The one above, besides providing the time, indicated the day of the month and the sun's position in the zodiac.
[Woodcut, 1567. © AKG, Paris]

pinpoint its precise position in the sky, say, right above the roof of an isolated house. You come back to visit a month later, at the same time: Orion is no longer above the roof, but next to it. It has shifted toward the west in relation to its previous position. Two months later, at the beginning of April, the situation has gotten even worse: at 9:00 P.M. Universal Time (UT), Orion is ready to set. The stars are not tuned to the same clock as the sun: they return to occupy the same position in the sky only every twenty-three hours and fifty-six minutes. This is the length of the so-called sidereal day, which corresponds to exactly one full rotation of the celestial sphere. Each time it completes a revolution, the sun has taken the time to "step back" several steps along the ecliptic in relation to its previous position. Rather than come back to the same spot in the sky every twenty-three hours and fifty-six minutes, the time it would take were it fixed to the celestial sphere, the sun always arrives a good four minutes later. Every twenty-four hours, that is.

2.
THE SECRET OF LONGITUDE

The clock, that brilliant invention for a world in which trains leave at 12:34 P.M. and scientists calculate the life spans of subatomic particles down to the billionth of a second, interested no one before the sixteenth century. All people needed was a mechanized reproduction of the sun's daily journey. One way to accomplish this is by having a hand complete one full rotation around a dial marked off into twenty-four sections in the same amount of time it takes the sun to mark a day and night around the earth. The time of day or night is indicated by the gradation to which the hand points on the dial.

A WAR OF POSITIONS

Sailors were the first to put their shoulder to the wheel of time, in their demand for precise clocks and ephemerides. These tools were required to help them determine their position at any moment and with reasonable precision. A ship's position corresponds to the point of intersection between the latitude and longitude of the parallel and meridian at which it is located. However, while finding their latitude posed sailors relatively few problems, finding their longitude proved to be a real headache.

The Bureau of Longitudes

The discovery and exploration of new continents, starting from the late fifteenth century, transformed the seas into an immense commercial and military battlefield. The great European powers would stop at nothing to control the maritime routes that led to the unknown and to colossal riches. To navigate effectively, the secret of the longitude had to be discovered at all costs. It's possible, in theory, to calculate the longitude by comparing the times at which various astronomical events occur (for example, the passage of a satellite of Jupiter, the onset of a solar eclipse, or the moon's position in relation to certain stars). In practice, though, you need very precise data concerning the movements of the stars and the moon in particular, which are usually far from evident. Astronomy soon became an affair of state: according to its first director, John Flamsteed, the Greenwich observatory was built on the order of King

Sailors had a tough time getting their bearings. They employed every astronomical means at their disposal—measurement of the positions of the stars, calculations of their movements and so forth—to orient themselves.

[Paris, BNF. © BNF]

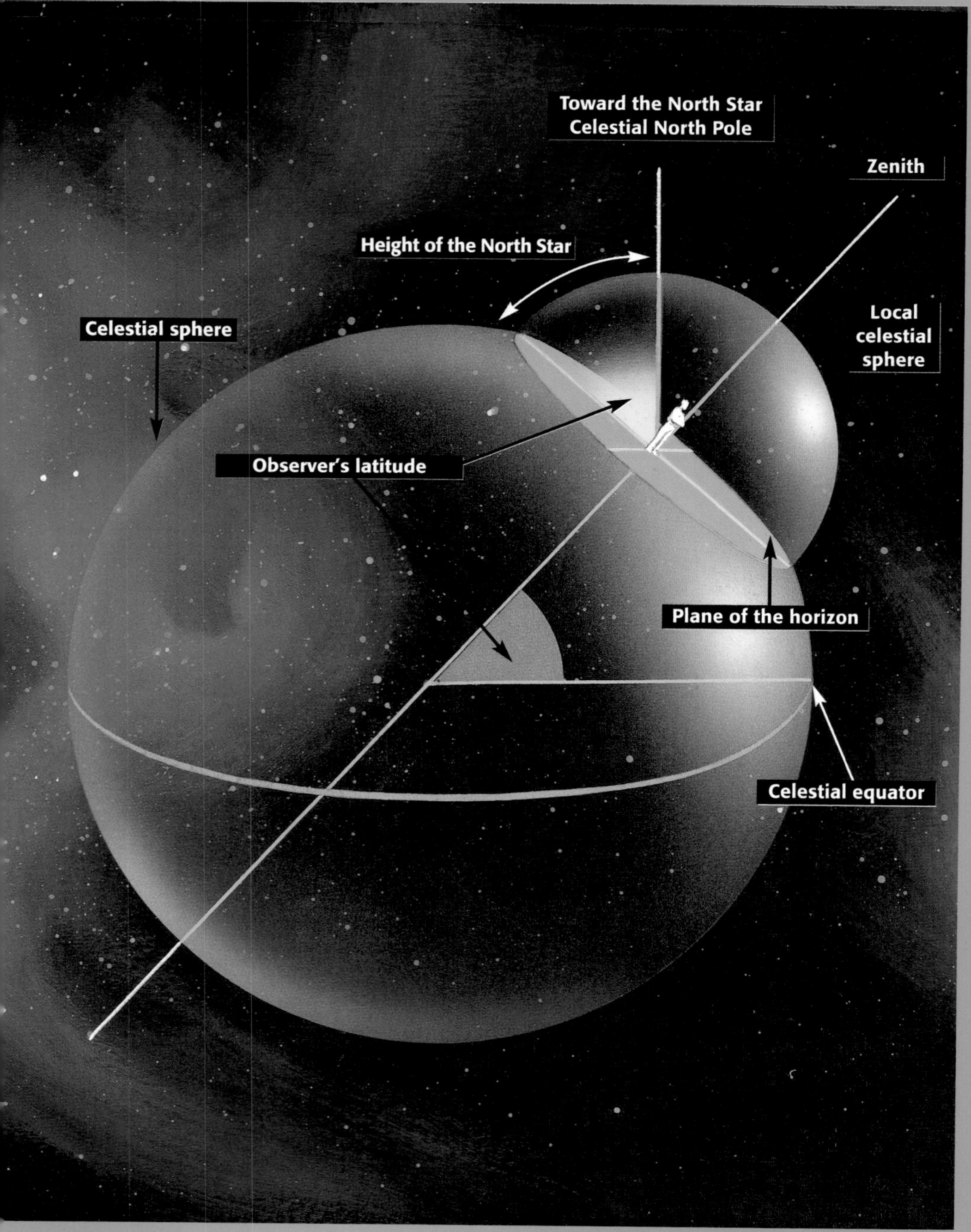

Toward the North Star
Celestial North Pole

Zenith

Height of the North Star

Local
celestial
sphere

Celestial sphere

Observer's latitude

Plane of the horizon

Celestial equator

TERRESTRIAL LATITUDES AND LONGITUDES

Latitude

Latitude is easy to find: it's the height of the North Pole of the celestial sphere in relation to the horizon. You need only measure the height of the North Star, with a slight adjustment to account for the fact it doesn't occupy the exact northern point of the celestial sphere (the correction to make is well known to astronomers).

Longitude

The sun, pulled along by the sphere of fixed stars, traces a complete circle parallel to the celestial equator every twenty-four hours. Now, a circle consists of 360 degrees. In an hour, the sun has the time to make an arc across the sky that corresponds to an angle of 15 degrees. The longitude is the angle, measured at the equator, which separates the prime meridian at Greenwich with that of, say, a ship. This angle can be deduced from the time it takes the sun to pass from the Greenwich meridian to that of the ship. To do this, you must first have on board a clock set to universal time, that is, to the Greenwich solar mean time. Let's imagine that at the moment the sun passes through the meridian of the ship, the clock reads 2:00 P.M. The nautical ephemerides give the hour, in universal time, of the sun's passage directly vertical to Greenwich. Let's say: 12:02. The difference in time, or one hour fifty-eight minutes, is then converted into angles. Knowing that the sun travels 15 degrees per hour, the longitude of our boat is 27°15' west.

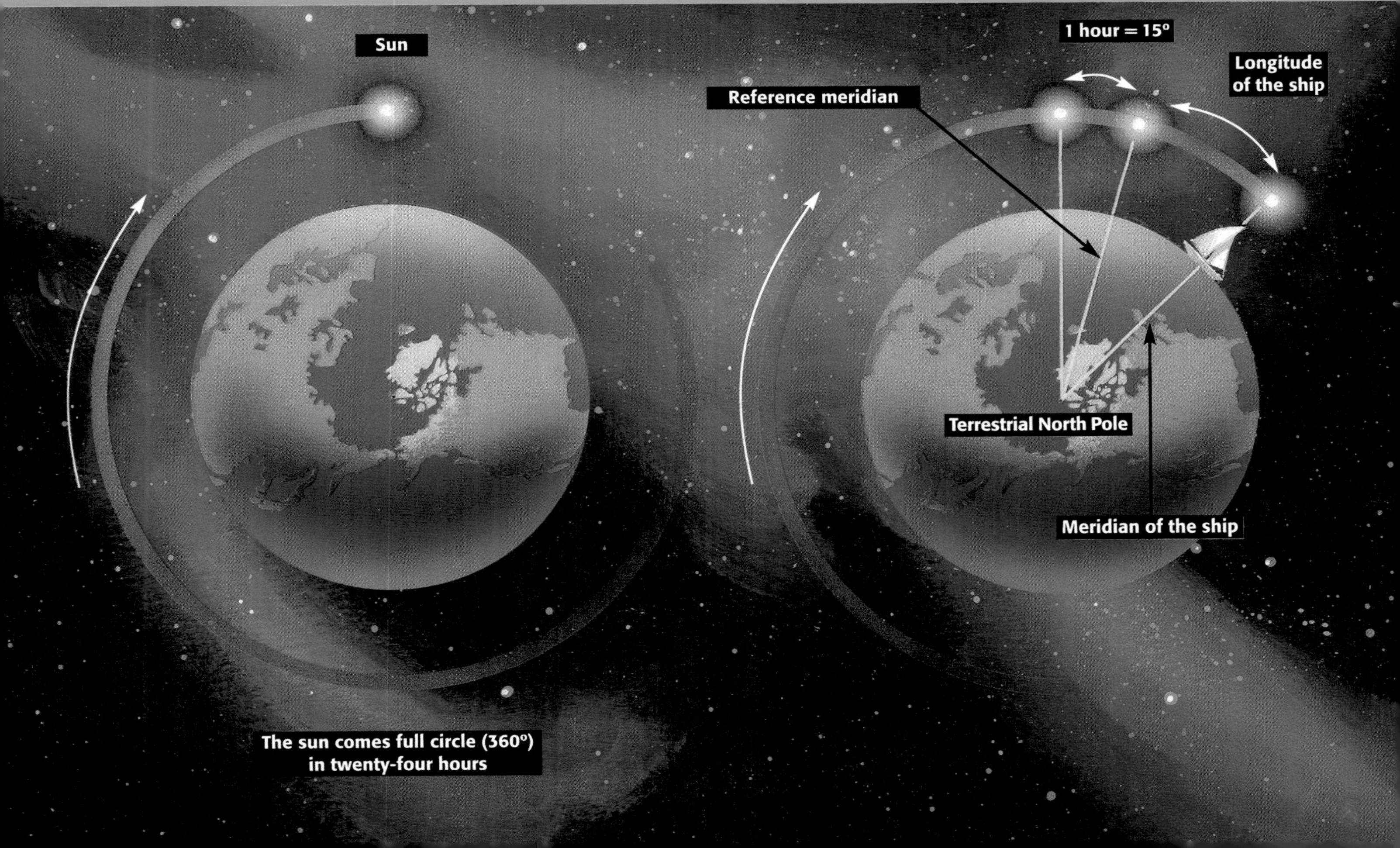

Sun

1 hour = 15°

Longitude of the ship

Reference meridian

Terrestrial North Pole

Meridian of the ship

The sun comes full circle (360°) in twenty-four hours

Charles II, who was horrified by a rumor that a mysterious Frenchman had discovered a way of calculating longitude by observing the movements of the moon. In France, a similar concern to equip sailors with the keys to the sky led to the creation of the Bureau of Longitudes on Messidor 7, Year 3 of the Revolutionary Calendar (that is, on 25 June 1795). England had already established a Bureau of Longitudes back in 1714, and the fledgling French Republic was convinced that the only means of upstaging that maritime power was to imitate her and invest in the stars. The Bureau, and later the Institute of Celestial Mechanics, was given among other missions that of developing anything that might facilitate and clarify the determination of longitudes, such as the implementation of new methods and the publication of ephemerides containing the astronomical parameters required by navigators.

One of the first marine chronometers, signed by George Graham (1735). These watches, carefully protected against rolling, accidents and humidity, were sometimes triple-locked in the captain's quarters and received high-level maintenance by the crew. Taking a bearing to determine the ship's longitude became a veritable ceremony, reserved for officers alone.
[London, British Museum. © The Bridgeman Art Library]

IF EVERY SAILOR IN THE WORLD . . .

The easiest method of finding the longitude involves comparing the time taken by the sun to move from a meridian chosen as a point of reference to the one of the ship (see sidebar, pp. 74–75). To stack the deck in her favor, England promised a reward of £20,000 to whoever succeeded in creating a clock capable of keeping precise time despite rolling, humidity and all the other little pitfalls of life at sea. In 1765, a certain John Harrison, carpenter by profession, hit the jackpot.

The Greenwich Meridian

Every nation referred to its own meridian for navigation. The French started counting from the one in Paris, the British began in Greenwich, while the Dutch referred to

a meridian crossing the Canary Islands. In 1884, the International Conference in Washington imposed upon the whole world the Greenwich meridian as the source for longitudes.

To the great joy of all those terrified by the astronomical and arithmetical acrobatics necessary to take a bearing at sea, navigators have happily learned to do without the stars for a good decade now. Simply become a proud owner of a transmitter-receiver connected to the GPS (Global Positioning System) navigation satellite network, and you'll know your geographical position with diabolical precision. Unfortunately, nothing's foolproof: if the system breaks down, you'd better be on good terms with the sky.

3.

FROM ONE TIME TO ANOTHER

Originally designed to replicate as faithfully as possible the daily movement of the sun, clocks wound up betraying our great star by revealing his unacceptable lack of punctuality. The daystar was rarely ever synchronous with clocks: upon its regular passage over the meridian, watches almost never read noon exactly. Making his daily rendezvous, the sun arrived either a few seconds early or a few seconds late. Chronometers designed to measure such deviations being paragons of exactitude, they were acquitted on all counts, leaving the sun to take the blame alone.

THE TRUE SOLAR DAY

This variation in the length of the day is the result of a curious phenomenon that had been observed as far back as the days of Mesopotamia: the sun doesn't move along the ecliptic at a constant speed. It sometimes slows down or, on the contrary, speeds up in its apparent orbit around the earth. These changes in rhythm remained unexplained until the seventeenth century, or after the earth had lost its status as the navel of the universe and began revolving around the sun in an elliptical trajectory (see chapter 6). It's still true today that when the sun slows down, the stretch of road it's traveling along the ecliptic gets shorter, and the day with it. Inversely, the day gets longer when the star speeds up. This elastic day is the true solar day.

THE MEAN SOLAR DAY

To account for these temporal fluctuations, we would have to develop clocks able to reproduce shifting time, where the duration of a second perpetually oscillates between two extreme values. What a pain for a science in full bloom at the time, but which required a fixed and perfectly determined standard for time both to measure the duration of natural and artificial phenomena and to speculate on the laws of the universe. Called upon for the umpteenth time to regularize the passage of time, astronomers were forced to abandon the sun as a point of reference. They calculated the length of a large number of true solar days, took their average and came up with a duration called the mean solar day. It's as if they had replaced the true sun with a fictitious sun, the mean sun, that rolls along the ecliptic at a constant speed. The duration of its daily journey around the earth, the mean solar day, doesn't vary a hair's breadth throughout the year, and it's this ideal day that was divided into twenty-four hours. The second was defined as one 86,400th of the mean solar day. Clocks were set according to this made-up sun, which always passes at zero hour in mean solar time over the meridian. Relative to the real sun, however, the mean sun always arrives a bit early or late. In February, it's vertical to the meridian a quarter of an hour before the sun is. In October and November, the opposite is true. The equation of time, which is equal to mean solar time minus true solar time, allows you to go from one to the other.

STANDARD TIME

The time problem was not yet solved as such. Legal time became mean time plus twelve hours: it's noon in legal time when the mean sun is vertical to the meridian. There are as many meridians as there are villages, towns, localities and cities on earth: time is local. For example, in France, when it is noon at the meridian in Paris, the sun has already passed the vertical of the one in Strasbourg, while lunch hour has not yet rung in Brest: the Bretons used to take their seats at the lunch table about fifty minutes after the Strasbourgeois did. Every city saw noon at different times. It was not until 1891 that France decided to simplify all that, by imposing upon the entire territory standard Paris time: when it's noon above the Cathedral at Notre-Dame, it's noon throughout the French territory. This is standard time. In England, time at the observatory in Greenwich has been in evidence since 1880. In the United States, railroads adopted standard time in 1883, although it wasn't officially adopted by the government until 1918.

UNIVERSAL TIME

In 1884, weary from all the temporal somersaults the countries were inflicting on each other, the whole world decided once and for all to set its pendulums at the same time, and created Universal Time, or UT: this is the standard time at the observatory in Greenwich. Standard time in all countries was now to be determined in relation to it. It was unfeasible, however, to insist that when it's noon in Greenwich it's noon everyplace else in the world: in one country people would be waking up at nightfall, while in others they'd be having lunch at dawn. The earth was therefore divided into twenty-four time zones. Within a single zone, time is the same for everyone and is equal to the one in Greenwich, plus or minus a whole number of hours (twelve being the maximum).

ATOMIC TIME

Is that it? Unfortunately, no. The modern world's unhealthy obsession with precision resulted in deep-sixing astronomical time based upon the earth's rotation.

The meridian line at the Paris observatory. For nearly two centuries it kept the rhythm of time for Parisians, who used to reset their watches to noon each time the sun passed overhead. The observatory remains the guardian of time, for today it houses the atomic clock.
[© *Ciel & Espace*/E. Graeff]

A Longer and Longer Day

Let's go over it one more time: the daily journey of the sun, the basis for chopping up time into days, hours, minutes and seconds, is only an illusion of the senses. In reality, the star is immobile, and it's our planet that revolves around it. The slightest modification to the earth's movement would have automatic repercussions on the apparent one of the sun, illusion or not. Drat! Double drat! Astronomers have largely proved that our proud earth doesn't spin smoothly. Its whirling movement around its axis is not uniform. It's subject to certain disturbances, one of which causes it to slow

down, continuously and ever so slowly. In other words, days are gradually getting longer by a few measly thousandths of a second per century.

The Cesium Second

These fluctuations, however, have been deemed significant enough that the time scale based upon the earth's rotation has been abandoned. Since the 1970s, the second has been defined as "the time interval equal to 9,192,631,770 periods of the radiation corresponding to the transition between the two hyperfine levels of the ground state of the cesium-133 atom." An atom oscillates like the pendulum of a large clock: it constantly shuttles between one state of energy and another. The number of oscillations it makes within a given period of time is remarkably constant, which explains why something so infinitely small was chosen as the new temporal scale. This is TAI, International Atomic Time. This doesn't mean UT is to be thrown overboard for fear of disconnection from the solar cycle, however. The TAI cannot waver from the UT by more than 0.9 of a second. When that occurs, one second is added on 31 December or 30 June at zero hour to UT, which has now become Coordinated Universal Time (UTC).

5

The Calendar

1. LUNAR TIME 2. TIME AND THE SEASONS 3. FROM JULIUS TO GREGORY

Seven-day weeks, 30-, 31-, 28- and 29-day months, years that leap and years that don't . . . Our calendar is far from simple and perfect. It is, however, the best system we have for breaking down long durations of time exceeding several days, and which we developed only after centuries and centuries of trial and error and ugly, arithmetic acrobatics. Its history is long, painful and relatively short on little sparks of astronomical genius, like the one that gave birth to the hour. And for good reason: what exists today represents a kind of shaky compromise between two different types of calendars. The first, the more ancient of the two, was based entirely upon the cycle of the moon and is the source behind our modern months. The second, born with agriculture, is governed by the sun and defines the year. Though these astronomical cycles, based on the moon and the sun respectively, have plainly nothing to do with one another, our great Mesopotamian, Egyptian, Greek and Roman forebears insisted on dividing time using both of them at once. The world became absurd: the return of spring was celebrated in the middle of winter, entire months disappeared mysteriously, the years got longer or shorter seemingly at random. Two reforms and the authority of an emperor and a pope, Julius Caesar and Gregory XII, were required before time could finally make sense.

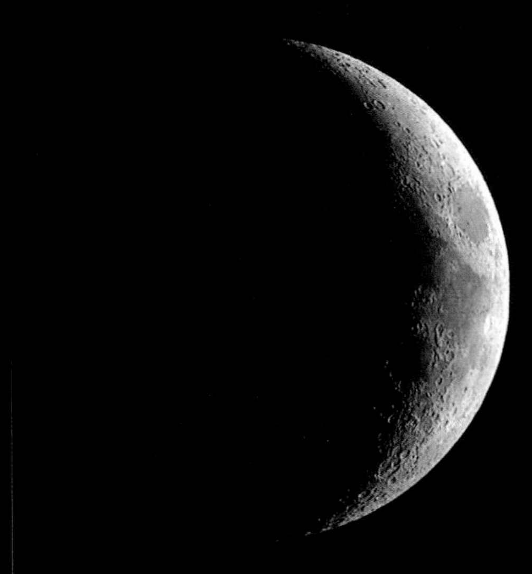

The moon is the earth's only satellite. It revolves around our planet at an average distance of 238,855 miles. It seems to have been born from a collision between the earth and another large celestial body. This supposedly tore off and ejected heavy fragments of rock into space. These rocks then compressed together to form the moon.

[© *Ciel & Espace*/A. Fujii]

1.

LUNAR TIME

To orient themselves over a period of time lasting longer than twenty-four hours, humans don't really have much choice. They can count the days separating them from a past or future event, such as a religious holiday, the way prisoners keep track of their sentence by making daily strokes on the wall of their cell. Long and tedious. People needed to find a way to locate themselves within the monotonous field of time by dividing it into clearly determined parcels. An astronomical cycle soon came around that for a long time did the trick: the cycle of the moon. This cycle repeats itself indefinitely and appears as obvious, natural and regular as the journey of the sun.

Following two pages: This splendid Turkish miniature, dating to 1583, depicts Ptolemy's world system. The twenty-eight days of the lunar month are spaced one by one over the zodiac; each box corresponds to one of the moon's phases.

[Map of the universe in *La Fine Fleur des histoires,* by Louqman. Istanbul, Istanbul Museum of Turkish and Islamic Art. © G. Dagli Orti]

BENEATH THE MOONLIGHT

The very least we can say about the moon is that she's constantly changing. One day she reveals her beautiful round and silvery face to the world. A week later, she shows nothing more than half her anatomy, before vanishing from the sky seven days later. She then reappears in the form of a very fine crescent, begins to grow again and, two weeks after her disappearance, has regained her broad, foolish face.

Lunation

This cycle, called lunation, lasts about twenty-nine days, and is due to the movement of the moon, the only body in the sky that actually revolves around the earth. The moon doesn't emit light; she only reflects that of the sun. Throughout her journey, she holds different positions relative to the sun and thus appears under constantly changing lighting conditions. Sometimes she's completely lit; at other times larger or smaller pieces of her remain in the shadows (see sidebar, pp. 86–87). To astronomers, lunation begins at the new moon, the moment when the satellite is invisible. It ends at the next new moon, at which point the next cycle begins without a moment's rest. Over and over again, lunations link themselves together into one gigantic, temporal garland.

LUNAR CALENDARS

These clusters of days occurring between two lunar absences were grouped into packets of ten or twelve in such a way that a second cycle was born: the year. It's longer, but it, too, repeats in time. To situate oneself within the year, each successive lunation was given a name. A calendar was born.

The lunar month has only one drawback. It doesn't contain a consistent or whole number of days. The time separating two consecutive new moons oscillates between twenty-nine days, six hours and twenty-nine days, twenty hours. On average, it's twenty-nine and a half days. Not knowing quite what to do with the extra half-day, users of lunar calendars chose to round it off: they made the months alternate between twenty-nine and thirty days over the course of their year.

In the Muslim calendar, which remains purely lunar, the year consists of twelve lunations of twenty-nine and thirty days. Day, according to Muslim tradition, begins at sunset. The first day of the month is the one on which the thin crescent following the new moon can be seen at sunset, about two days after the moon's disappearance. If, due to cloudy weather, say, the crescent cannot be seen, the beginning of the month is put off for a day. This type of calendar requires particularly clement skies, at the risk of watching the months stretch out indefinitely.

A Year in Rome

In Rome, a government time official was charged with keeping a lookout for the appearance of the new moon's crescent and proclaiming the *calends*, or the beginning of the month. This was a holiday, like the *ides* which fell in the middle of the month, either thirteen or fifteen days after the *calends* (the fourteenth was considered bad luck). The *nones* occurred the ninth day before the *ides*. The primitive Roman year seems to have contained only ten months. The first, March, was dedicated to the god Mars; the second, April, was supposedly that of buds. Maia, goddess of Growth,

The moon revolves around the earth. There's nothing virtual about this movement. The moment she moves directly between the sun and the earth, she's said to be "in conjunction" and is invisible. Our favorite night satellite doesn't emit light, but reflects that of the sun; at the moment of the conjunction, her luminous face is turned toward the sun, and we can only see her darkened half. This is the new moon. She continues her route, shifting further

THE PHASES OF THE MOON

and further away from her initial alignment. She now begins to reveal her lit side a bit. A thin crescent, which thickens as the days pass, remains visible later and later into the night. About seven days after the conjunction, half the moon's side is lit: it's the first quarter. The illuminated portion continues to grow, passing through the phase of the gibbous moon before presenting us her face, now big, round and inundated with sunlight.

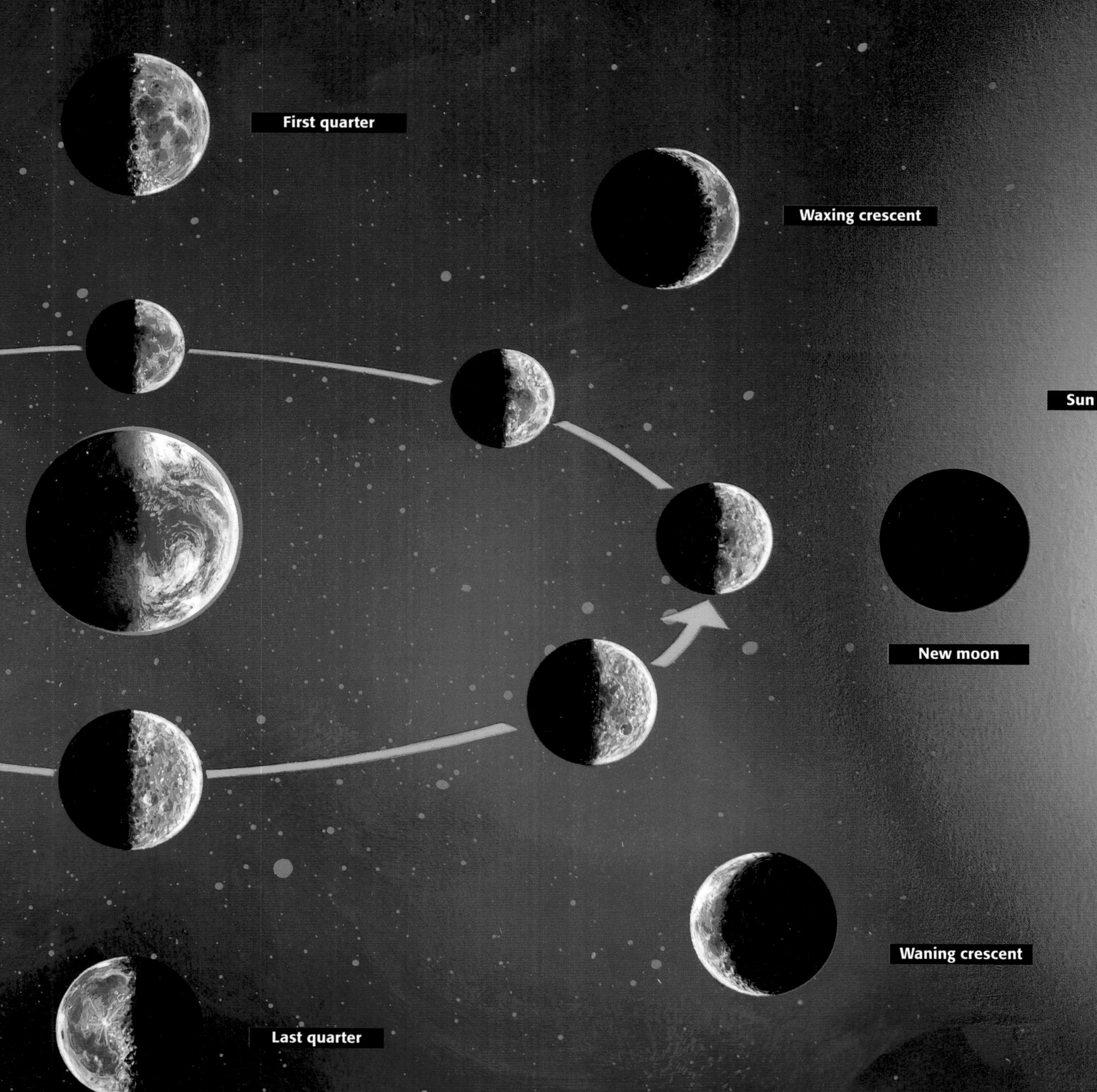
First quarter
Waxing crescent
Sun
New moon
Waning crescent
Last quarter

This is the full moon, which occurs about fifteen days after the new moon. She continues her course, drifting toward the east. Each day, she hides a little bit more of her silvery face from us. The moon passes back through the gibbous phase until, seven days later, she again reveals only half of her side. This is the last quarter, visible throughout the latter half of the night. The moon continues to shrink, becoming a crescent again before slipping between us and the sun. This is the new moon.

The duration of the cycle of lunar phases is the period of time elapsed between two oppositions. That is, on average, 29.5 days (29 days, 12 hours, 44 minutes, 2.8 seconds). This is longer than the sidereal month, the time taken by the moon to reoccupy the same position relative to the stars, and that measures an average of 27.3 days (that is, 27 days, 7 hours, 43 minutes, 11.5 seconds). During this little month, the earth, revolving around the sun, has had the time to travel a short distance in its orbit. To regain the same configuration, such as a conjunction, the moon needs to catch up with the earth, which takes more than two days.

watched over the third, May, while Juno governed the fourth, June. Then, in a sudden lack of inspiration, the Romans named the last six months according to their rank: *Quintiles* was the fifth, followed by *Sextilis*, the sixth, then *September, October, November* and *December.* The year was subsequently extended by two months: January, dedicated to the god Janus, and February. Just goes to show how far back it all goes.

2.

TIME AND THE SEASONS

In all regions where agriculture—invented more than twenty-three thousand years ago—became humankind's principal resource, the passage of time was given a particular rhythm by the succession of seasons, schematized to the extreme. There was first the time for plowing, then the time for sowing and then the time for harvesting. This cycle, like that of the moon, repeated itself endlessly: crops were born in spring, grew in the summer, ripened in the fall and died in the winter.

THE SEASONAL YEAR

Every plant, whether cultivated for food, like wheat, or to satisfy other needs, like cotton or flax, has its own unique qualities: some need to be planted in the middle of spring, others pruned at the end of autumn or planted out in winter. The farmer had an interest in making the right move at the right time, or risk losing all. Farmers needed time markers whose appearances or metamorphoses always coincided with the beginning of each season.

The Sun

A complete cycle—spring, summer, autumn, winter—begins and ends on the day of the spring equinox. This is merely a result of the sun's apparent annual movement around the earth. The beginnings of each season, that is, the spring and fall equinoxes and the winter and summer solstices, coincide with the moment the sun occupies a particular position along the ecliptic (see chapter 2). The time separating two consecutive returns of the sun to the vernal point, the spot on the ecliptic it holds on the spring equinox, is called the tropical year. Today, this equals 365.2422 days, or 365 days, five hours and approximately forty-nine minutes. Lacking the ability to determine the sun's position through direct observation, it's still possible to deduce it using related phenomena, such as the spot on the horizon where the sun rises (see sidebar, p. 69) or the variation in length of a gnomon's shadow (see sidebar, p. 91).

TIME AND THE STARS

The stars and constellations make much better points of reference. The sky's appearance changes sharply from one season to the next (see sidebar, p. 92). Handsome Orion, for example, is a winter constellation. He's in best form during the nights of January and February, and boycotts those of July and August. The opposite is true for Cygnus, Lyra and Aquila, who reveal themselves in all their splendor only in summer and hibernate in winter. As a general rule, the constellations easily visible at midnight on winter nights are still present in the celestial vault in summer, but only smack at noon, and vice versa.

A beautiful sixteenth-century illumination depicting the earth, the planets and the zodiac. The outer planetary sphere is surrounded by several circles representing, in order: the months, the succession of days, the figures of the zodiac, their names, their symbols and the dates on which the sun can be found in them.

[Naples, Biblioteca nazionale Vittorio Emanuele III]

Sirius is the most brilliant star in the sky. Its appearance before dawn marked the beginning of the Egyptian harvest year. It belongs to the constellation Canis Major (literally, the "Great Dog," at center left), who trots for eternity behind his master, Orion (on the top right).
[© *Ciel & Espace*/A. Fujii]

Heliacal Rise

Our ancestors were very attentive to the so-called heliacal rising of the stars. The sun constantly moves along the ecliptic, shifting in relation to the sphere of fixed stars. From time to time it occupies a position such that an unfortunate star nearby disappears from the sky. A star, located in close proximity either right above or below the sun, will rise along with it. A few days later, the sun will have moved eastward. It has liberated the stretch of sky that contained the star and everything looks as if the sun, in relation to it, has stepped back several steps along the celestial sphere. The star now rises just before the sun does: this is what astronomers call "heliacal rise."

Tomorrow, at Dawn . . .

The reappearance of visible stars in the wee hours of the dawn coincided with the return of good or bad weather, and served as a sign to farmers. The Greeks used almanacs that associated various agricultural labors to the heliacal rise of certain stars and constellations. Hesiod's poem, *Works and Days,* dating from the eighth century B.C., provides one famous example: "When the piercing power and sultry heat of the sun abate, and almighty Zeus sends the autumn rains, and men's flesh comes to feel far easier, for then the star Sirius passes over the heads of men, who are born to misery, only a little while by day and takes greater share of night, then, when it showers its leaves to the ground and stops sprouting, the wood you cut with your axe is least liable to become worm-eaten [. . .]. But when Orion and Sirius are come into mid-heaven, and rosy-fingered Dawn sees Arcturus, then cut off all the grape-clusters, Perseus, and bring them home." One star, moreover, owes its name to the agricultural task to which it has been linked: Vendemiatrix, in the constellation Virgo, derives from the Latin *vindemiator,* or "grape-picker." Its heliacal rise occurs in late-August/ early-September, right around harvest time.

THE EGYPTIAN CALENDAR

Sirius's heliacal rise, in mid-July, announced the beginning of the great summer heat, earning it the following treatment by Virgil: "the well-known brightness of Sirius, bringing thirst and diseases to weary mortals, rises and saddens the sky with [its] baneful light." The Egyptians, in contrast, venerated it.

The Myth of the Distant One

According to Egyptian mythology, the eye of Ra, the only part of the sun god's anatomy that we can see, is also his daughter. Heads-side up, she's Hathor, the beautiful and golden goddess of Love. Tails-side up, she's Ra's evil eye, the terrible Sekhmet, a fearsome goddess with the muzzle of a lion. According to the Myth of the Distant One, Ra's daughter went off to live far away from both her father and Egypt, on the deserted edges of the world. She was a fearsome lioness who devoured anyone who came near her. Pining for his daughter's return, Ra sent the gods Shu and Thot to look for her. To avoid provoking the goddess's wrath, they took the form of tiny harmless monkeys. Then Thot began to turn on the charm. Pouring flattery after flattery upon her, he slowly succeeded in winning her over and convinced her to follow them home. She grew calmer and calmer, eventually regaining the beautiful face of Hathor upon her rearrival in Egypt. Her return was triumphant: life came back along with her.

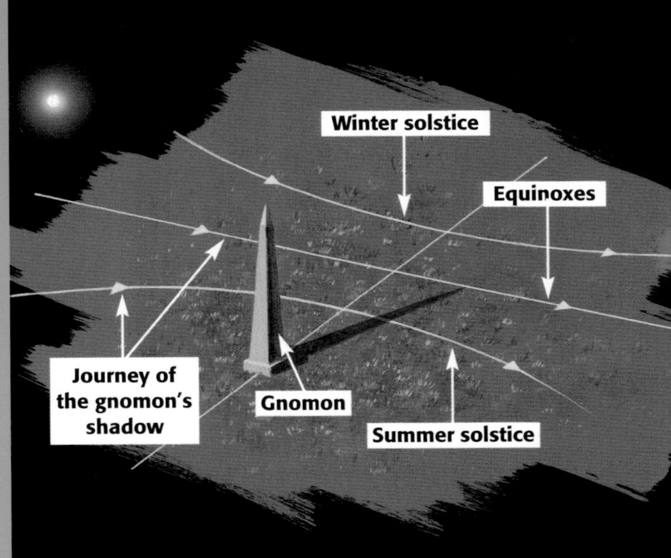

THE GNOMON

A simple rod or giant obelisk planted vertically in the ground can act as a gnomon. The gnomon's shadow is at its longest at the winter solstice, when the sun is at its lowest point along the southern horizon. It's shortest at the summer solstice, when the sun is high above the horizon. Its size is halfway between these two extremes on the days of the two equinoxes.

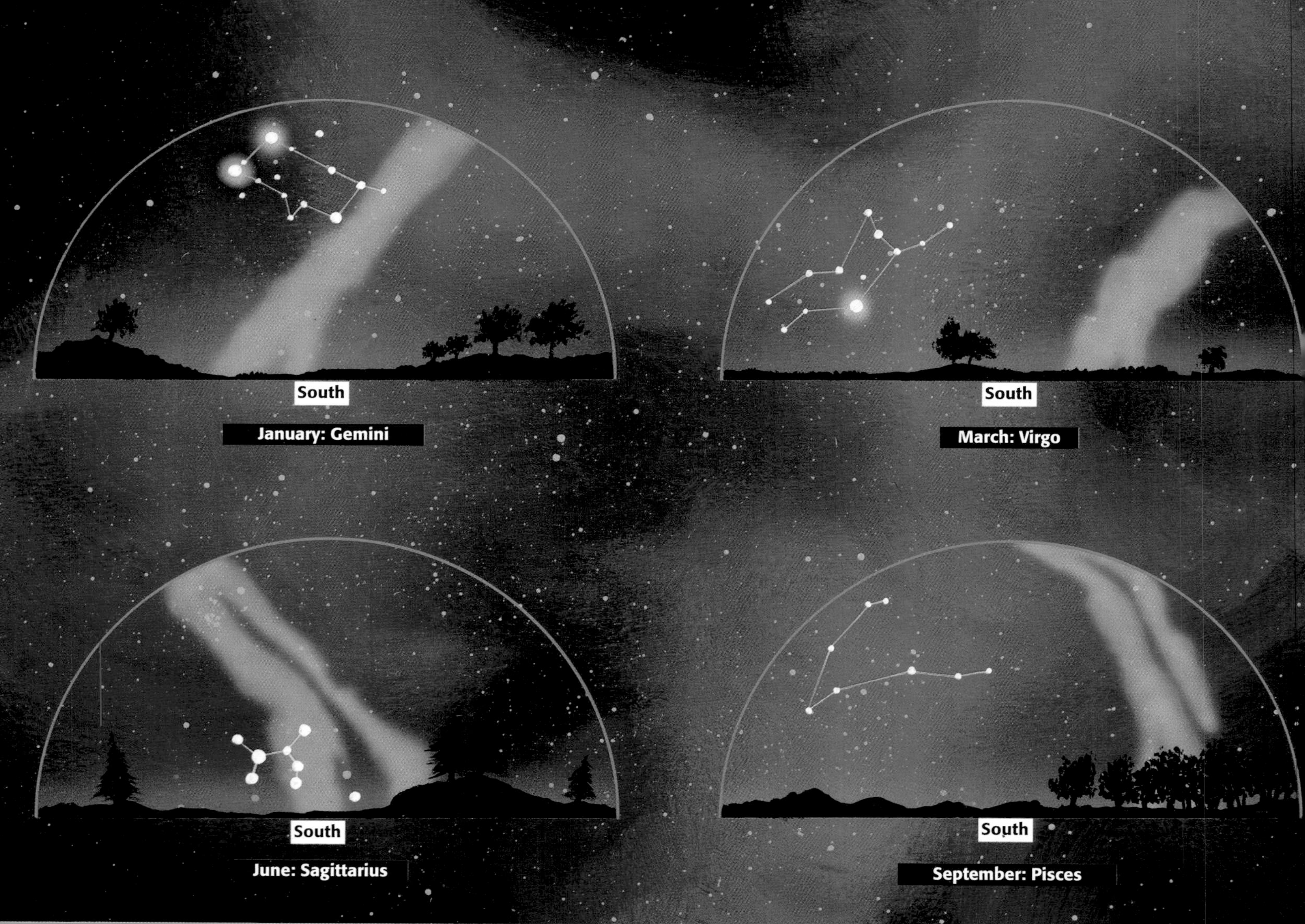

SEASONS AND CONSTELLATIONS

The ecliptic, the annual route of the sun, is marked out by the constellations of the zodiac. The daystar moves through one to another. In January, the sun can be found in the constellation Sagittarius, which rises and sets with it and is absent from the nocturnal sky. Gemini, occupying the diametrically opposite spot on the ecliptic, are clearly visible. In March, the sun has moved and can be found in Pisces. It's Virgo who profits from this. In June, Sagittarius takes his revenge on Gemini, who now contain the daystar. In September, it's Virgo's turn to get pinched, leaving the advantage to Pisces.

The Flooding of the Nile

This legend is intimately related to an event that was capital for the Egyptians: the return of the annual flooding of the Nile, whose waters covered arable lands from the end of July through late October. Without this, life would have deserted the land of Egypt. The days leading up to the flood were especially painful: the lioness Sekhmet was on a rampage, while the level of the Nile was at its lowest, leaving its stagnant waters favorable to epidemics. In the sky, meanwhile, Ra beat down mercilessly. It was only around 20 July that the star Sothis, our Sirius, at last deigned to put an end to the Egyptians' anguish. Its heliacal rise announced the Distant One's return to the fold, flooding was nigh, and the people, relieved, let themselves go in endless rejoicing. Once again, Ra had provided for them.

The god Thot is sometimes depicted in the form of a baboon, but other times he has the beautiful head of an ibis. The patron of scribes, this cunning diplomat regularly had to settle the problems that might spring up among the heirs of the sun god Ra. As the guardian of the moon, his magic enabled the moon to continually recover her silvery body.

[Bas-relief, Thebes, Valley of the Kings. © The Bridgeman Art Library]

The Vague Year

Several millennia before that time, Sirius's heliacal rise and the onset of the annual flood both occurred on the first day of the Egyptian year. This extraordinary coincidence supposedly drove the Egyptians to modify their calendar. The year consisted of three seasons for them: Akhet, the season of the flood, then Peret, that of vegetation, followed by Shemu, that of the harvest. Each was in turn divided into four thirty-day months. In total, the primitive Egyptian year had only 360 days. Five more were added, called epagomenal days, causing New Year's Day, or the first day of the first month Thot, to coincide with the reappearance of Sirius. This was a dreadful mistake.

The Egyptians surely must have calculated the time separating two consecutive heliacal rises of Sirius: this period doesn't equal 365 days, but 365 and a quarter days. Let's imagine, then, that in a given year the star's morning appearance falls right on the first of Thot. After four years, this will occur on the second of Thot. Then on the third, fourth, fifth, and so on. After 120 years, Sirius will rise one month after New Year and the great festivals marking his return. The Egyptians stuck with this system, called the "vague year," however, for a good number of centuries, until the Romans forced them to adopt the Julian calendar.

3.

FROM JULIUS TO GREGORY

There could be far worse systems than the Egyptian calendar. The Greeks, Romans and Mesopotamians, for their part, got tangled up in an absurd system combining seasons and lunations. They insisted on retaining a year made up of twelve lunar cycles. Now, by alternating months of twenty-nine and thirty days, this year runs about eleven days too short compared to the seasonal year. Imagine that the first day of a year containing twelve lunar months coincides with that of the spring equinox. Three years later, the latter will land at the beginning of the second month of the year, then the third after six years, the fourth after nine years, and so on. The equinox does not return to the same date; it roams throughout the calendar. It won't coincide again with the New Year for another thirty years. This is why Muslim holidays constantly change in relation to the Western calendar, moving ahead ten or eleven days each year.

In order that the lunar year stick more or less to the cycle of the sun, so that the equinox always falls at least within the same month, it was necessary to add an additional month to it regularly, every three years. In ancient Rome, the priests were responsible for defining the duration of this thirteenth month, named Mercedonius, and the moment action needed to be taken to realign the calendar with the seasons. They did so only at their whim, using their power instead to extend the duration of their magistrate friends' terms of office, or to advance or hold back as they pleased the day taxes were due.

Julius Caesar's Reform

Weary of such caprices, Julius Caesar summoned from Egypt a Greek astronomer, Sosigenes, whom he ordered to straighten out the mess. Sosigenes began by assigning a definitive length to the year, specifically, 365.25 days. He shortened it to 365 days, which he divided into eleven months of 30 and 31 days, plus one numbering only 28—February. Dropping the digits after the decimal point was out of the question, however, lest the calendar start rambling off again. To catch up with its drift, rather, Sosigenes resorted to a clever trick: add an extra day to the year once every four years. Thus were born leap years, in which February contains 29 days. New Year's Day, which used to fall on 1 March, was replaced by 1 January in order that, according to Sosigenes's calculations, the spring equinox falls on 25 March. This calendar, called Julian, became effective on 1 January, A.D. 45.

THE GREGORIAN CALENDAR

In A.D. 325, on the orders of Emperor Constantine, Church elders gathered together at the first Council of Nicaea to debate crucial questions surrounding Christianity. They took advantage of the occasion to fix the date of Easter once and for all, which was to be established in relation to the spring equinox. That year's equinox didn't occur on 25 March, however, but on the twenty-first. Poor Sosigenes was accused of having erred in his calculations, and the council adopted 21 March as the definitive date of the equinox.

Sosigenes's Error

Sosigenes made two mistakes. The spring equinox occurs once every 365.2422 days, the length of the tropical year, and not every 365.25 days. The Julian year is too long by 0.0078 days. After four hundred years, spring arrived three days ahead of its intended 25 March arrival. Meanwhile, even the initial schedule was wrong: the equinox didn't actually take place on 25 March in year A.D. 45, but on the twenty-third or twenty-fourth. This is why it fell on 21 March in the year of the Council.

Years and the Precession of the Equinoxes

Sosigenes must have confused the tropical year with the sidereal year. Being longer, the latter corresponds to the period of time taken by the sun to return to a given position relative to a star. The time separating two consecutive revolutions of the sun to and from the celestial meridian of Sirius, for example, is 365.256 days. Hipparchus was aware of the discrepancy between these two kinds of years because of the precession of the equinoxes (see chapter 3). In his era, the sun entered the constellation Aries right on the spring equinox.

To fix the date of Easter, defined by the Council of Nicaea as "the Sunday following the fourteenth day of the moon coming of age on 21 March or immediately thereafter," the ecclesiasts relied upon dry, arithmetic gymnastics called *computus*.

[*Psalter of Saint Louis and Blanche de Castille*, thirteenth century. Paris, BNF. © BNF]

Today, it's in Pisces at that moment and doesn't join Aries until the month of April. As the centuries go by, it will arrive later and later to its stellar rendezvous. The wonderful coincidence of Sirius's heliacal rise—which happens nowadays in August—with the flooding of the Nile was slowly eroded by this phenomenon.

Pope Gregory XIII's Reform

In 1582, the equinox took place on 11 March. The Julian calendar had to be corrected immediately to return Easter to its proper date. To do so, Pope Gregory XIII assembled an ad hoc committee, which proposed relatively simple solutions. First, the equinox had to occur on 21 March. Gregory XIII shortened year 1582 by ten days: inhabitants of Rome went to bed on Thursday, 4 October and woke up on Wednesday the fifteenth. Next, aware that every four hundred years contain three days too many, they deleted the extra days. The secular years, those which end in '00—1400, 1500, 2000, 2300, and so on—are all leap years in the Julian calendar. There are four of them every four hundred years: by making three of them "normal," the extra days disappear. Only one of these secular years remains a leap year, the one divisible by four hundred. The year 2000, for example, was a leap year, and so will be 2400 and 2800. But not 2100, 2200, 2300, 2500 or 2600. The very Catholic Spain and Portugal went along with this reform at the same time as the Vatican. France opted to remove the ten-day surplus in December 1682. Protestant and Greek Orthodox countries, in the words of astronomer Johannes Kepler, "preferred to disagree with the sun than agree with the Pope." England and Sweden didn't adopt the Gregorian calendar until 1752, Russia until 1918, Romania until 1919 and Turkey until 1924.

6
The Planets

1. THE MOVEMENTS OF THE PLANETS
2. UNIVERSAL SYSTEMS
3. KEPLER'S THREE LAWS

If astronomy had cared about the sun and stars alone, the earth might have persisted in taking itself for the navel of the universe indefinitely. The sphere of fixed stars did an honest job meeting the needs of cartographers and government time officials. Its regular movement and smooth shapes let beautiful harmony reign throughout the cosmos. Harmony, alas! Ruined by the planets. Their capricious movements, so contrary to icy spherical order, proved to be a thorn in astronomers' sides from the fifth century A.D. to the present. The planets categorically refused to submit to the arbitrary laws upon which the Greeks, from Plato to Ptolemy, built their cosmic edifices. The Greeks did everything they could to coerce these rebel bodies into respecting the image of the world they wanted to believe, a universe in which everything revolves around a cosmic center, the earth. Yet try as they might to rein the planets in

The nine planets of the solar system. Mercury, Venus, Earth and Mars form a separate group, those of the telluric planets. Made of solid materials, they are all practically the same size. Of the five others, Jupiter, Saturn, Uranus, and Neptune are gaseous giants that measure four to eleven times the size of Earth.
[© *Ciel & Espace*/JPL]

using a multitude of circular orbits, the latter nonetheless continued to behave as they pleased and almost never appeared where astronomers expected them to be. The situation began to improve when the very discreet Nicolaus Copernicus, in the mid sixteenth century, got the rich idea of banishing the earth from the center of the universe and replaced it with the sun. Johannes Kepler then caught the rebound and, after years of relentless work, discovered the key to the movement of the planets. The face of the universe had completely flipped around: the sun now held the place of honor, while the planets, including the earth, revolved around the sun following orbits in the shape of ellipses.

1. THE MOVEMENTS OF THE PLANETS

The Greeks were only aware of five planets, the ones visible to the naked eye: Mercury, Venus, Mars, Jupiter and Saturn. The too faint and distant Uranus, Neptune and Pluto wouldn't be discovered until well after the invention of the telescope: the first in 1781, the second in 1846 and the last in 1930. "Planet" means "wanderer" in Greek: brighter and more colorful than the stars, these bodies move in paths nothing at all like the peaceful rounds of the fixed stars.

The Celestial Highway

Observed from the earth, the movement of the planets resembles that of the sun. Like it, they drift in relation to the stars. They are not fixed upon the celestial sphere, and appear to be moving in enormous circles around us. A longer and deeper analysis of their apparent movements begs the conclusion that they all follow in the wake of the sun: the planets never wander far from the ecliptic. They all remain within the sort of starry belt formed by the zodiac, which resembles a two-lane road divided by the white stripe of the ecliptic. The planets all roll along in the same eastward direction. They zigzag a bit, sometimes passing above or below the ecliptic, but such deviations are never severe. They each follow the road at their own pace: Mars, for example, takes about two years to come full circle around the earth, while Saturn stretches it out for over thirty years.

Retrograde Movement

This is where the planets' contribution to the harmony of the spherical edifice ends. Giving them a closer look, their trajectories don't appear linear like those of the sun and moon—which the Greeks tried to toss in the same bag as the planets—but looped. Their behavior is rather mind-boggling. They move along perfectly normally across the celestial sphere when, suddenly, they stop as if they'd just come up with a clever idea and need time to reflect. They then make an about-face and start heading in the other direction. They quickly change their minds and stop a second time, make another about-face and continue on their "normal" route. As astronomers put it, they retrograde regularly in their apparent orbits, backpedaling a bit before returning to the straight and narrow, and follow trajectories shaped like loops, S's and Z's across the sky (see sidebar, pp. 106–107).

Variable Distances

The distance of each planet to the earth varies over time. The apparent diameter of Mars, for example, doesn't remain constant. Its glowing, red disc appears larger and brighter at certain times of the year than at others. Unless you believe it's a planet of variable geometry, inflating and deflating, fattening and slimming down to the rhythm of some insane cosmic diet, the Mars phenomenon only makes sense if you consider the planet as moving sometimes toward, sometimes away from the earth. When it's further away, it appears smaller and fainter than when it's closer.

The Inequality of the Seasons

Finally, the planets don't travel their apparent orbits at a steady speed. They have a tendency to slow down when they're crossing certain notches on the zodiacal belt—the ancients called this phenomenon zodiacal preference—and to accelerate within others. The sun exhibits a similar behavior, which is at the root of the inequality of the seasons. Spring, summer, fall and winter are not equal in length. According to calculations made by the Institute of Celestial Mechanics in France, in 1998 winter lasted 89 days, spring 92 days, 18 hours, summer 93 days, 15 hours and autumn 89 days, 21 hours. Everything acts as if the sun moves more slowly in its orbit in summer than in winter.

Evening Planets, Morning Planets

Naturally, this entire little world was confined to the interior of the sphere of fixed stars, the ultimate barrier between Everything and Nothing. The planets, though

Mars became very fashionable in the nineteenth century, thanks to the canals astronomers thought they saw on its surface. Everyone thought they were the work of Martians, made famous in 1897 by H. G. Wells in *The War of the Worlds,* in which villainous octopuses invade the earth. [© *Ciel & Espace*/NASA/MSSS]

they zigzag in their paths around the earth, are still subject to the same law that governs the sun and stars: diurnal movement. The planets make their daily trip along the horizon from east to west, as if they, too, were being dragged along by the celestial sphere. When their journey leads them to hold the same position as the sun relative to the stars, they rise and set with the daystar and disappear at night. Mercury and Venus literally ride the sun's coattails. Some days you'll find them a few paces behind and only visible at sunset, other days they're ahead of the sun and shine in the sky just before dawn.

Mathematical Acrobatics

The Mesopotamians were aware of every little planetary and solar quirk. With one eye on the heavens and the other on the *Enuma Anu Enlil,* their great book of divination collecting several thousand omens and directions for their use, they became very familiar with the celestial bodies and the eccentricities of their movements. Thanks to the zodiac and the ecliptic, which provided them a system of reference for keeping track, the Mesopotamians acquired a taste for prediction around the fifth century B.C. They relied on arithmetical recipes they'd figured out based on their innumerable observations. For example, to determine the length of the day on which the sun will be found in a precise spot within the constellation Taurus, you have to perform a certain number of operations using very specific figures: begin by subtracting ten from X, multiply by Z and then add A. But, when the sun's in Leo, these figures won't be the same at all: you have to subtract ten from Y, then multiply this by N, then add B . . .

2.

UNIVERSAL SYSTEMS

Babylonian empiricism, with its eccentric accounting of the movements in the sky, was surely not the Greeks' cup of tea. Holding their noses a bit, the Greeks did manage to salvage one aspect of the Babylonian baby from the bathwater—astrology—and remodeled it in light of their own philosophical and scientific understanding of the world.

Rather than merely keep track of celestial movements—to make do, basically—in the fifth century B.C. they began trying to replicate such movements and determine their underlying mechanism using all the geometrical resources available to them at the time. This was roughly like attempting to visualize the wheels of a clock by observing only the movement of its hands and without the possibility of peering inside. The Greeks strove to develop models of the universe, theoretical constructions they could count on to predict the positions of celestial bodies without having to resort to observation.

KEEPING UP APPEARANCES

The man who initiated this vast program, if we're to believe fifth-century philosopher Simplicius, who himself was only repeating earlier statements, was the philosopher Plato. He who had transformed the universe into a gigantic sphere spinning around the earth (see chapter 1) supposedly also established the credo by which all astronomy would swear for twenty centuries to come: We must keep up appearances! There is by

Plato's Three Commandments

Careful, however—you can't envision the sky any old way—Plato established very precise specifications for those who wish to uncover the secrets of the universe:

1) The earth must remain still and at the center of a spherical universe.

2) You may only explain the trajectories of celestial bodies as combinations of circular movements.

3) The speed of every heavenly body must remain constant across the celestial sphere.

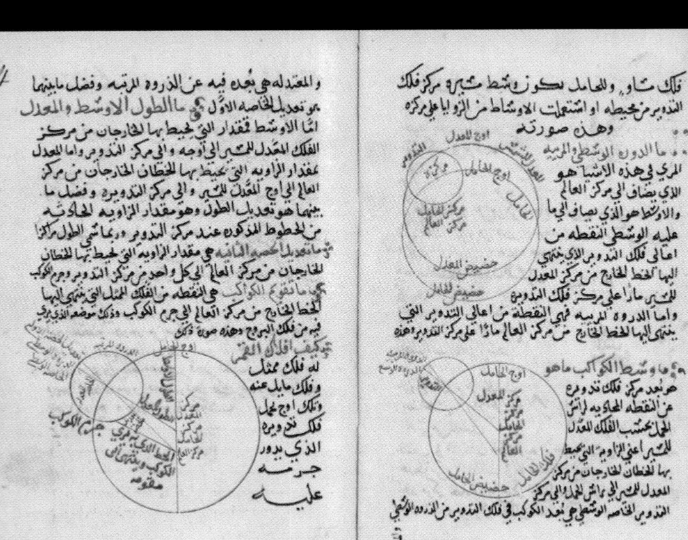

Excerpt from a seventeenth-century astronomy manual based on an earlier work by the Arab astronomer Al-Biruni (973–1050). He replicated the orbits of the planets following Ptolemy's system. Reconciling Ptolemy's vision with Aristotle's homogenous and concentric spheres posed many problems to heirs of the Greeks.
[Paris, BNF. © BNF]

THE GEOCENTRIC SYSTEM

The Platonic restrictions, which rested upon sound philosophical argument, were pretty frightening. Imagine, first of all, that the planets travel in perfectly circular and steady orbits around the earth. This is a perfectly respectable rule. Observed phenomena, however, must then be ignored: like the stars, the planets will have to maintain a constant speed and distance from the earth and avoid making loops in the sky. Deviations from these ideal motions were considered anomalies by the Greeks, who did everything they could to explain them away and keep the earth at the center of the cosmos. The set of solutions they proposed make up what is called the geocentric world system.

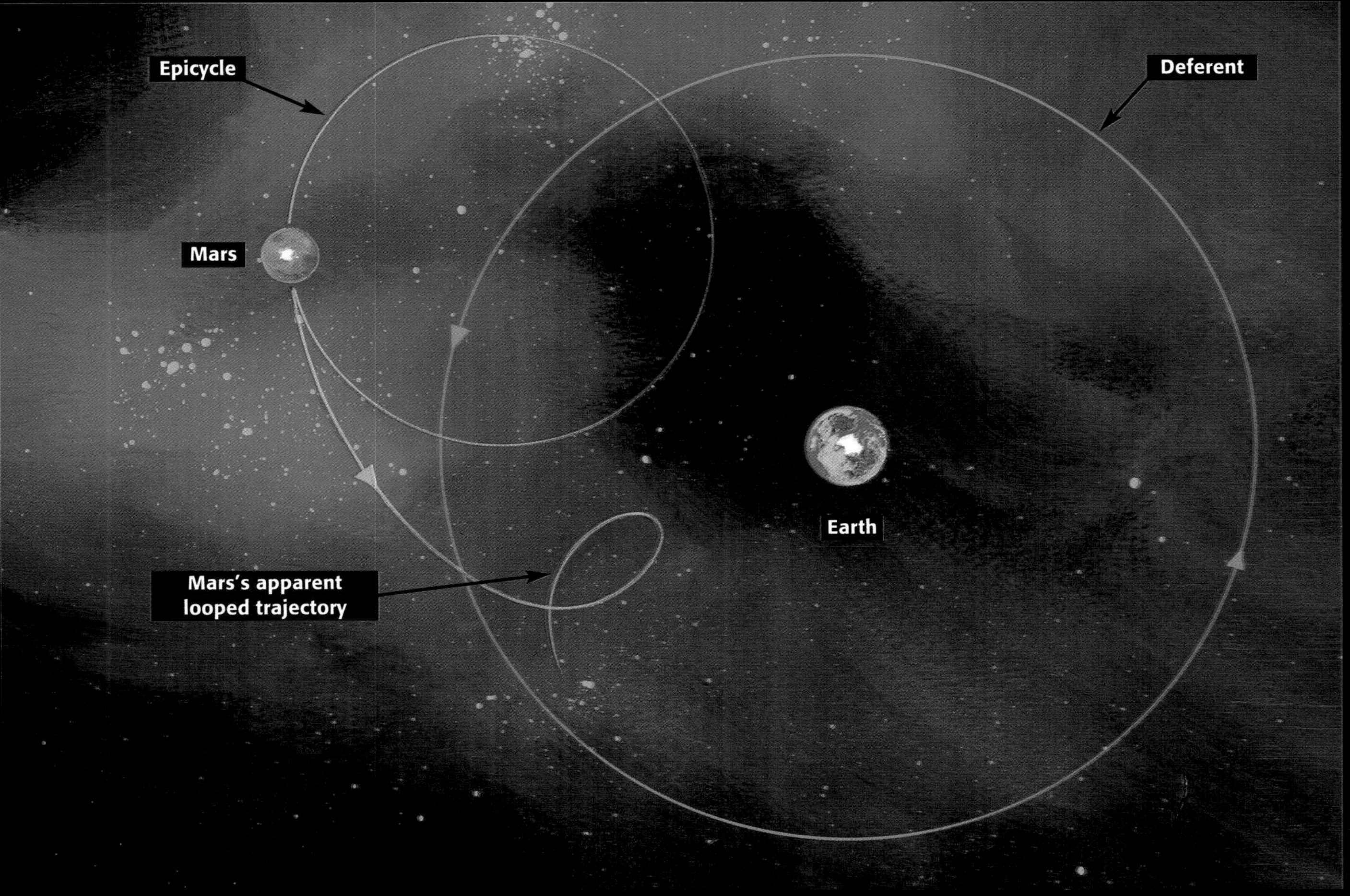

To explain Mars's retrograde movement, Ptolemy resorted to artifice: the planet, he argued, makes a little circle—an epicycle—the center of which travels around the earth. To an observer, Mars appears to make a regular about-face in the sky, and its trajectory is looped.

The Eccentrics

To explain the inequality of the seasons, Hipparchus proposed that the sun moves with constant speed but in a circular orbit not directly centered on the earth. The star's distance to the earth would then vary throughout the year. The closer the sun got to us, the more it would appear to slow down. The further away it was, the faster it would seem. This type of off-centered orbit is called an *eccentric*.

MARS'S RETROGRADE MOVEMENTS

Mars belongs to the so-called superior planets, located outside the earth's orbit. These planets don't present phases like Venus and Mercury and can be observed at different times at night. They are in conjunction when they are aligned with the earth and on the opposite side of the sun. This is when they are at their furthest point from us; their apparent diameter is minimal and they will only become visible again in the early morning a few days later. They are in opposition when the earth slips between them and the sun. This is the ideal time to observe them, for they are large, very bright and visible all night long. Finally, they are in quadrature when the angle they make with the earth and sun measures 90 degrees. They will only appear during part of the night.

Mars's zigzag trajectory is a pure illusion, due to its difference in speed with the earth in their respective orbits. The illusion resulted from the fact that passengers on the earth (human beings, that is) were completely unaware of the movement and shape of our celestial vehicle's orbit. The two planets are like two trains

The different positions of Mars

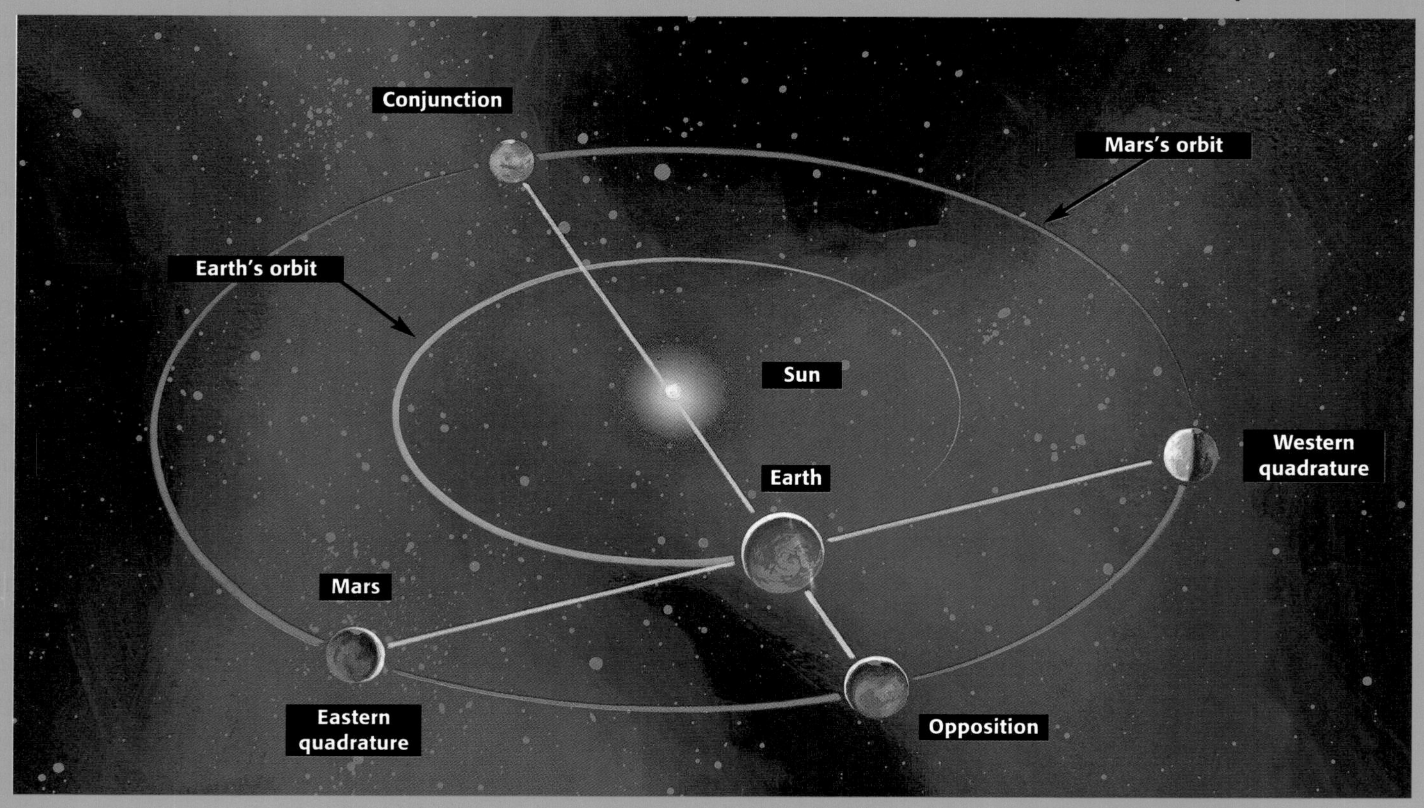

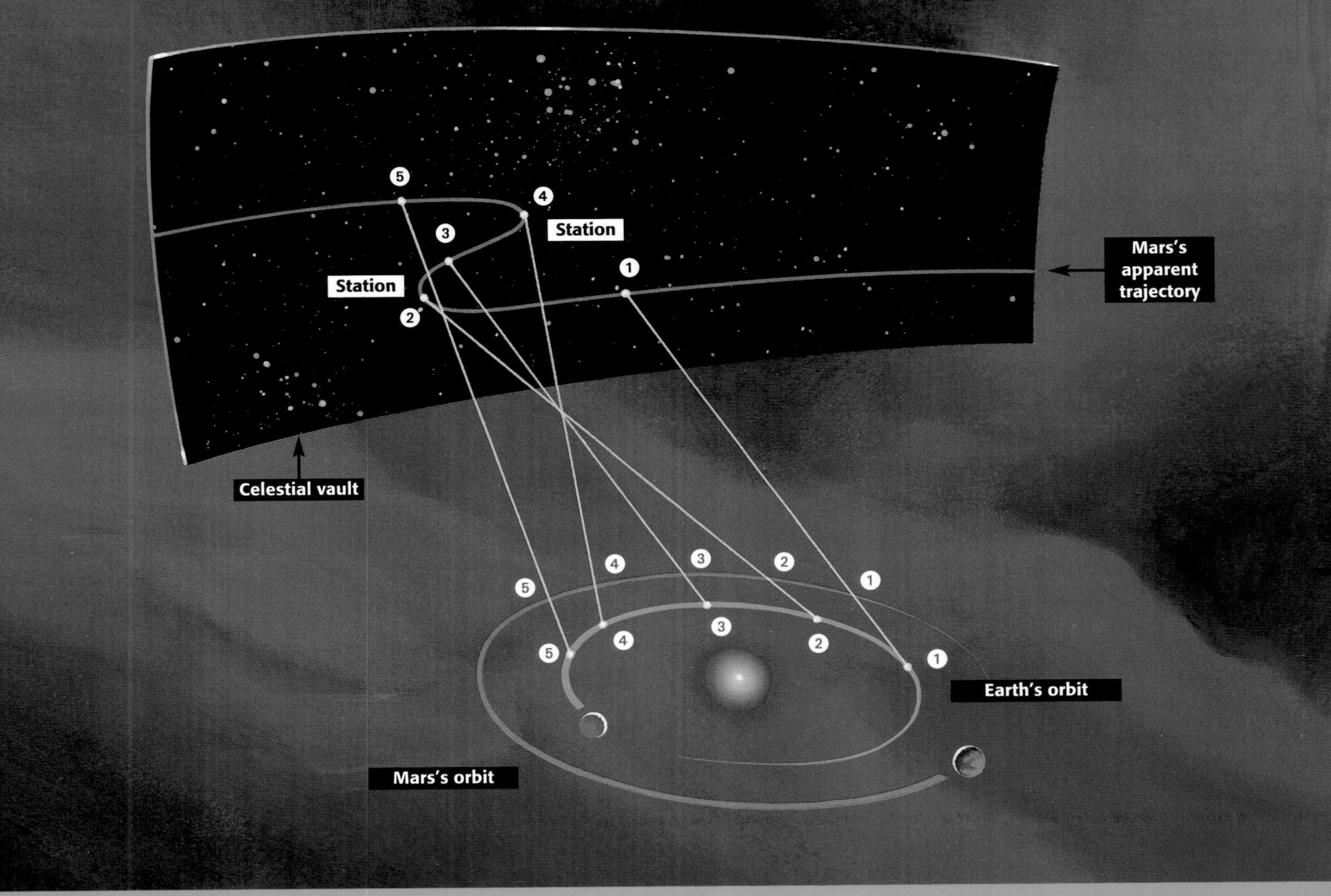

Mars's retrograde movements

rolling in the same direction on an oval-shaped circuit. Each moves on its own track. Let's imagine that Mars, which has a certain lead on the earth, goes around the first turn. We earthlings, convinced we are stationary, see the red planet moving straight ahead in its path. We then see it hesitate and come to a halt, before making an about-face toward us. It reaches our level and passes us by, and we now have to turn around to be able to see it. It goes round the second bend backwards, seems to stop a second time, and then begins its route in the right direction again. In reality, the earth is faster than Mars (which takes 1.88 years to complete its orbit); we're the ones who actually took the first turn, caught up and passed the red planet, before making the second turn.

The stern canon Nicolaus Copernicus. No revolutionary was more discreet than he, who held off publishing his heliocentric model of the solar system until the very last minute. The story goes that Copernicus received the first printed copy of his *On the Revolutions* the day he died.
[Krakow, Collegium Maius. © AKG/E. Lessing, Paris]

Epicycles

This doesn't solve the problem of Venus's great variations in distance and brightness, for example, nor of Mars's mysterious retrograde movements. Willing to consider any combination of celestial movements provided they be circular and uniform, the mathematician Apollonius of Perga in the third century A.D. came up with the *epicycle:* this is a little circular movement made with constant speed by the planets and whose center travels around the earth along a wider circle, the *deferent*. The deferent may itself be eccentric, provided it settles matters for astronomers (see p. 105).

The Equant

Eccentrics and epicycles were not enough to explain the seemingly complex movement of Venus. Ptolemy must have had to seriously twist Plato's dogmas around in inventing the *equant*. This is a point located above the center of the deferent in relation to which the speed of an epicycle's center remains constant. Plato was clear, however: the movement has to be smooth and regular in relation to the center of the circle.

The Long Reign of Ptolemy

These combinations of circular movements and his little equant trick, clearly spelled out by Ptolemy in his master work, the Almagest, remained the astronomer's only tool until the seventeenth century. Little did it matter if they conformed or not to reality. Perhaps they were merely a mathematical artifice, a purely theoretical reenactment of planetary trajectories. But who cared, provided they kept up appearances? The Greek provided all the necessary data—the earth's position in relation to the equants, the time taken by each planet to complete

its epicycle, the duration of the epicycle's orbit—to calculate the different positions the planets may hold at different moments. The exactness of these measurements was sometimes lacking, and the planets sometimes arrived too early, too late or never to the rendezvous astronomers had with them. But for want of a better system, everybody made do, and Ptolemy's calculations were used for centuries after his death.

THE HELIOCENTRIC SYSTEM

Errors were teeming in the astronomical tables of the era, but this is not what drove the quiet Polish canon, Nicolaus Copernicus, to formulate a new world system in the first half of the sixteenth century. According to his own writings, rather, he was disgusted by the equant and the twisting which had been inflicted upon Plato's rule of uniform circular motion. Fashioning himself as the new Ptolemy, Copernicus decided to start all over again from scratch, this time replacing the earth with the sun. What philosophers and historians of science later called the Copernican Revolution was, in the end, merely the application of the corkscrew principle to the ancient celestial sphere.

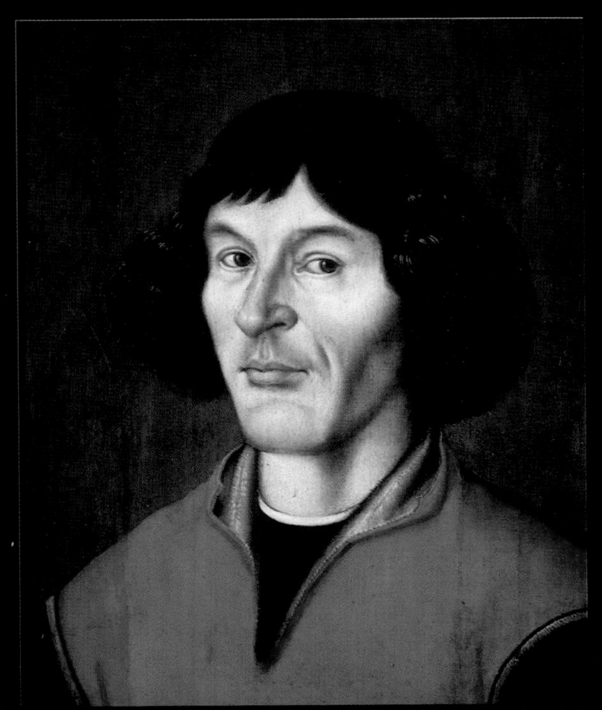

Nicolaus Copernicus

There was absolutely nothing revolutionary about Nicolaus Copernicus. He was born in 1473 in Thorn, or Turun, a Polish village located along the Vistula. At age twelve, after his father's death, he was placed in the custody of his uncle, Lucas Watzenrode, bishop of Varmia, who later helped him land a cushy job in the Church by having him nominated canon of Frauenburg in 1497. Before taking this position, Copernicus had gone to Krakow to study in 1491, and then moved on

Excerpt from a fifteenth century manuscript depicting the image of the universe according to the Greek Heraclides of Pontus (fourth century B.C.).

[Marciana Capella, *De Nuptiis Philologiae et Mercurii,* fifteenth century. Venice, Biblioteca nazionale Marciana]

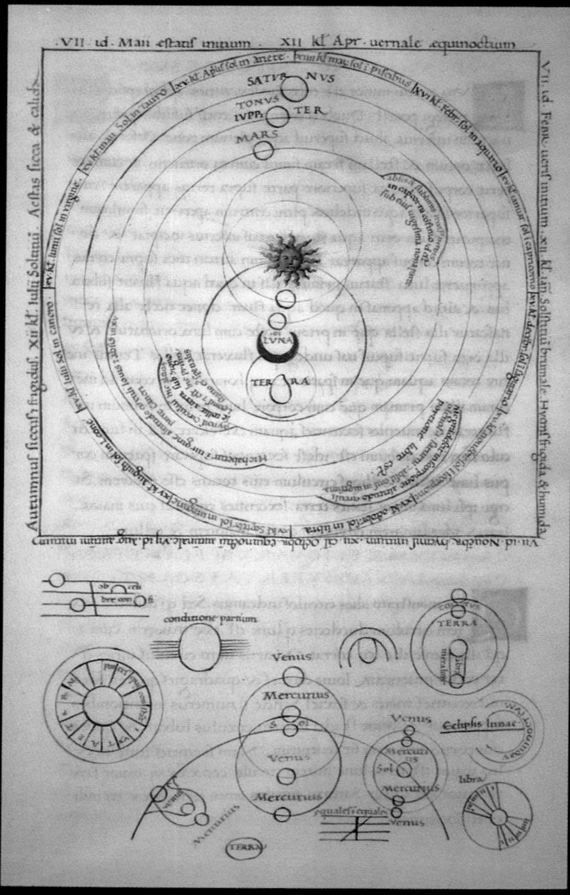

to Bologna where he was introduced to astronomy. After Bologna, he went to Rome, Padua, and finally Ferrara, where he earned his diploma in canonical law in 1503. He left to spend a few years working as his powerful uncle's doctor before he moved definitively to Frauenburg. He was treated generously by the Church, whose vast estates it was his duty to manage and maintain. In this peaceful, opulent monotony, Copernicus quietly polished his astronomical theory and composed his only major work, *On the Revolutions of the Heavenly Bodies.* He circulated several abstracts among trusted friends, who begged him, in vain, to publish it. A young and fiery mathematician, Joachim Rheticus, became his lone disciple in 1539. Rheticus wrested permission from his master to publish the *Narratio prima* in 1540, in which some of his ideas were presented. But Copernicus's book itself did not appear until 1543, the year of the canon's death.

From the Earth to the Sun

In Copernicus's heliocentric model, the sun takes the place that belonged to the earth in previous systems. All the planets known at the time—in order: Mercury, Venus, Earth, Mars, Jupiter and Saturn—revolve around it. Only the moon continues traveling around the ex-navel of the universe which, along with its movement around the daystar, is rotating on its own axis. For Copernicus, the diurnal rotation of the stars is only apparent and the sphere of fixed stars is . . . fixed. The idea wasn't new. Heraclides of Pontus, in the fourth century B.C., had already applied the corkscrew principle to the celestial marble: have the earth whirl around or have the stars do the spinning, both ways look the same to an observer on the earth. Heraclides also suggested that Venus and Mercury move around the sun, which offers an elegant explanation for their constant proximity to the daystar.

Heraclides's model had no more success than that of Aristarchus of Samos, a contemporary of Plato, which was exactly the same as Copernicus's.

Always Circles

This model also provided a definitive answer to the problem of looped orbits: retrograde movements aren't real, but an effect of perspective due to the motion of the earth (see sidebar, pp. 106–107). On the mathematical side, however, it didn't do much to simplify life for astronomers. Copernicus kept an iron grip on Plato's smooth, circular orbits, which the celestial bodies continued to follow in his model. The center of the universe wasn't exactly the sun, but the navel of the earth's own trajectory. To keep up appearances and explain away variations of speed and distance, his system required a number of epicycles and eccentrics almost as great as those in Ptolemy's system (the equant, however, was history). One small step for mankind . . .

The heliocentric model of the solar system as conceived by Copernicus. He retained the notion of celestial spheres, including the one of the fixed stars. It still encloses the universe, even if the rotation of the earth around itself renders it useless.

[Nicolaus Copernicus, *De revolutionibus orbium coelestium,* libri VI, 1543. Paris, BNF. © BNF]

3.

KEPLER'S THREE LAWS

Copernicus's book, which was particularly dry, didn't make much of a stir among the era's astronomical bigwigs. Heliocentrism appeared much like some excellent but vague idea, tossed out at the end of a meal by a guest who's just left the table. By luck it was overheard

Posthumous portrait of Johannes Kepler as imperial mathematician. He compared himself to a fierce and stubborn dog, always dependent on others, ready to accept the lowliest jobs, happy to gnaw on bones and detesting baths. He was also a pious and tolerant man, who drew up his own horoscope from time to time. He predicted, moreover, that his marriage would be calamitous.

[Munich, Deutsches Museum. © AKG, Paris]

not by a deaf man, but by a myope with double vision: Johannes Kepler.

"POOR" KEPLER

Kepler's childhood was depressing. He was born in 1571 to a Protestant family in Württemberg, in Weil der Stadt near Stuttgart. He was the eldest son of a soldier named Heinrich Kepler, a drunken and bad-tempered mercenary. Sickly and fragile, young Kepler was treated hardly better than a dog in a familial house stuffed with uncles, aunts, cousins and grandparents. He went to school for a little while, but quit when his parents bought an inn with an appropriate name: The Golden Sun. His epileptic brother set fire to the inn and his father disappeared for good. Fortunately, Johannes was pulled out of the gutter thanks to a scholarship offered to poor children of merit by the Dukes of Württemberg. He entered the seminary, and then the University of Tübingen, where he studied theology with the aim of becoming a pastor. His teacher, Michael Maestlin, introduced him to heliocentrism, of which he soon became a fervent supporter.

The Divine Beauty of the Universe

Kepler didn't finish his studies and left for Graz, in Austria, where a job as a mathematician was waiting for him. He was a strange fellow, at the same time fiery and timid, arrogant and obsequious, complaining and quick-tempered. He might, in sudden and inexplicable changes of mind, bite the hand he'd been kissing the day before and hurl abuse on a man he admired. He was also intelligent, sensitive, curious and equipped with a certain sense of self-mockery which kept him from taking himself too seriously. Convinced that the God to whom he'd sworn his existence could only have created a world of beauty and intelligence, he spent his life seeking both these qualities, both in the sky and on earth. The ideas of Copernicus, who never offered even a shadow of concrete proof to support his system, seduced Kepler because they restored a certain harmony to the cosmos. The planets are arranged in order of decreasing speed, from the fastest, Mercury, which completes its orbit in about one hundred days, to the slowest, Saturn, which takes around thirty years. Johannes Kepler was also a veritable heliolater, a sun worshipper, who found it much more logical to place the bright star at the center of the system instead of nasty old earth. He suspected a mystical correspondence between the heliocentric cosmos and the Holy Trinity: the sun held the place of the Father, the sphere of fixed stars that of the Son, and space, which he filled with a sort of ether, that of the Holy Spirit. But why six planets? Why do they follow these trajectories and not other ones? Why these speeds in particular? For Kepler, chance had nothing to do with it. God had a plan. The Creator designed the world the way a geometer would have, and there is certainly a mathematical, geometrical link between all these orbits.

Tycho Brahe's Legacy

Kepler offered his services to the Prince of the Stars, Tycho Brahe, who accepted. The Dane, in staunch disagreement with Copernicus, had concocted his own vision of the world: the planets revolved around the sun, which moves around an immobile earth. Brahe had carefully studied the movement of Mars, and wanted to use his observations of the red planet, the best of the era, to prove the validity of his system. To do so, he needed Kepler's mathematical gifts. In February 1600,

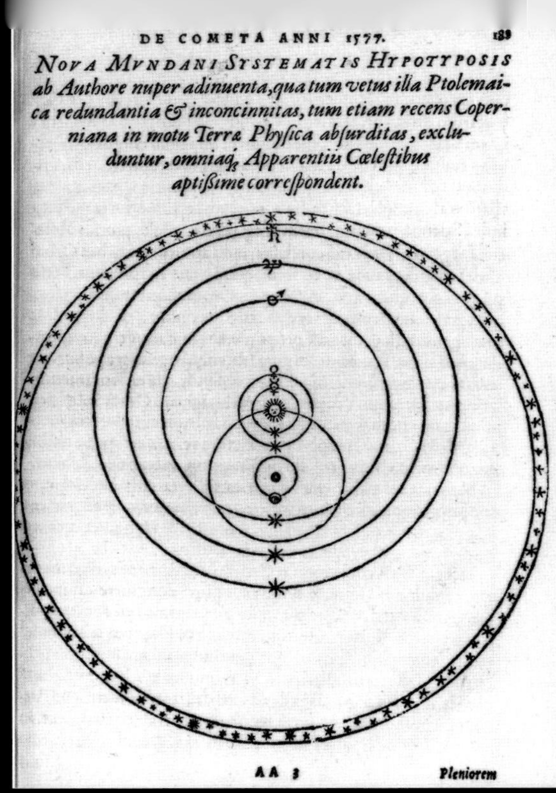

DE COMETA ANNI 1577.
189
NOVA MUNDANI SYSTEMATIS HYPOTYPOSIS ab Authore nuper adinuenta, qua tum vetus illa Ptolemaica redundantia & inconcinnitas, tum etiam recens Coperniana in motu Terræ Physica absurditas, excluduntur, omniaq; Apparentiis Cœlestibus aptissime correspondent.
AA 3
Pleniorem

Tycho Brahe's system of the universe, 1588. Jesuit scholars, who refused to accept heliocentrism and knew the flaws of Ptolemy's model, nevertheless had a weakness for him.
[Paris, BNF. © BNF]

KEPLER'S THREE LAWS

FIRST LAW

This is the easiest one: the trajectory of the planets follows an ellipse. Don't imagine it like an egg, but rather like a slightly stretched out balloon. An ellipse contains two axes of symmetry: the shorter is the minor axis, cut at right angles by the longer one, the major axis. It also contains two foci, points located on the major axis and symmetrical in relation to the minor axis. The sun is not at the center of this ellipse, but occupies one of the foci. The closest point to it in the ellipse is the perihelion, and farthest from it the aphelion.

Eccentricity is another characteristic of an ellipse. It corresponds, in mathematical language, to the distance between one of the foci of the ellipse and its center, divided by the value of half the major axis. Concretely, the eccentricity gives you an idea of the shape of the orbit: if it's worth zero, the center and focus blend together, and we're dealing with a circle. The closer the value of the eccentricity approaches one, the greater tendency the focus has to approach the edge of the ellipse, and the flatter the ellipse becomes.

SECOND LAW

The planet-sun radius vector covers equal areas in equal time. The surface swept by a wire stretched between the sun and a planet over a given amount of time will always be the same, no matter where it's located in its orbit. This implies that, the further away the planet is from the sun, the less elevated its speed. The closer it is, the faster it is. The speed is maximal at the perihelion and minimal at the aphelion.

THIRD LAW

The square of the length of time it takes a planet to make a full revolution around the sun is proportional to the cube of the major axis of the ellipse of its orbit. Whichever planet one considers and the size of its orbit, the ratio of the square of the orbit period to the cube of the major axis is a number that remains constant throughout the solar system. The time taken by a planet to make one revolution around the sun, its own "year," depends completely on the size of its orbit. And vice versa: the orbit period cannot be touched without changing the ellipse.

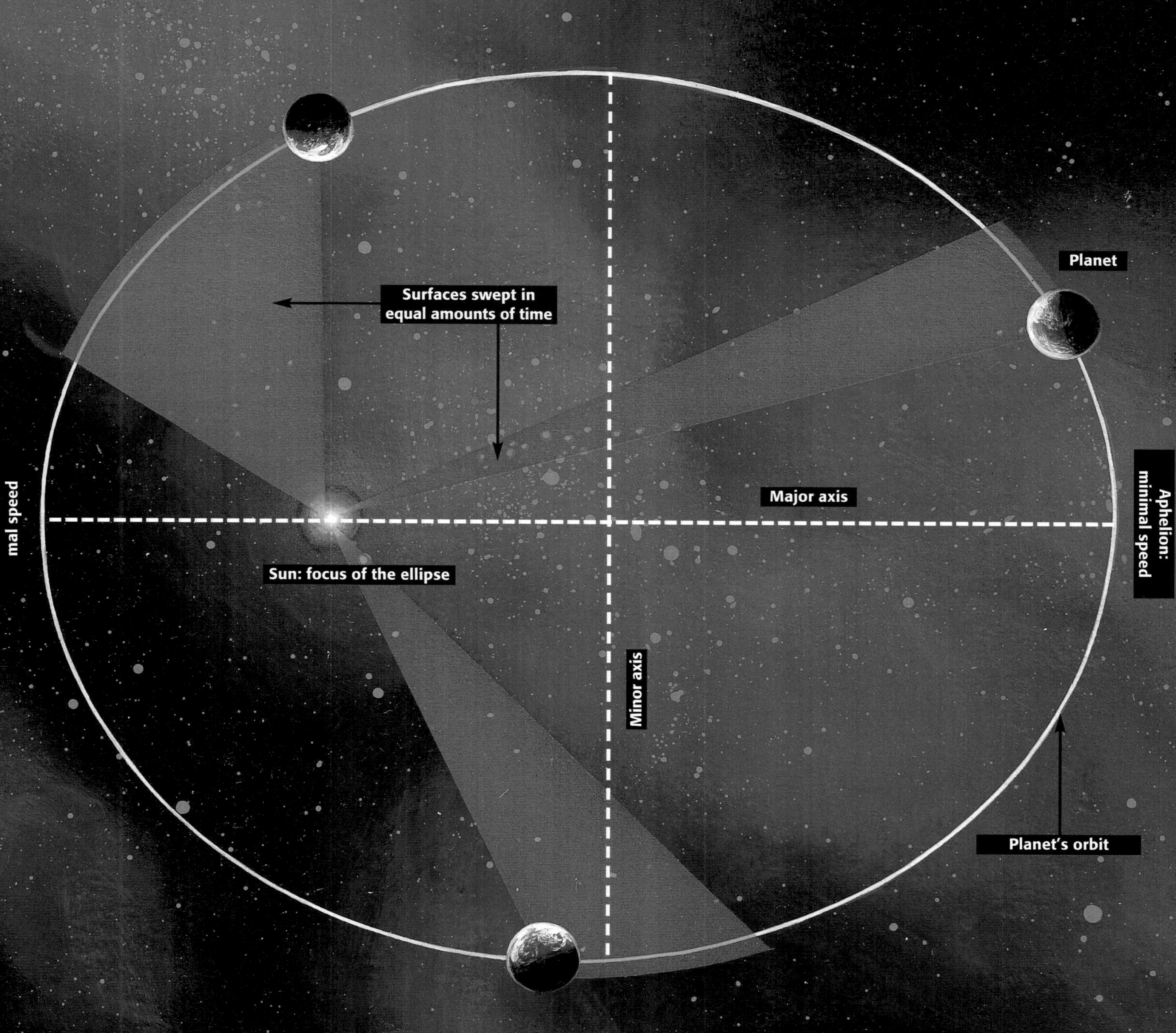

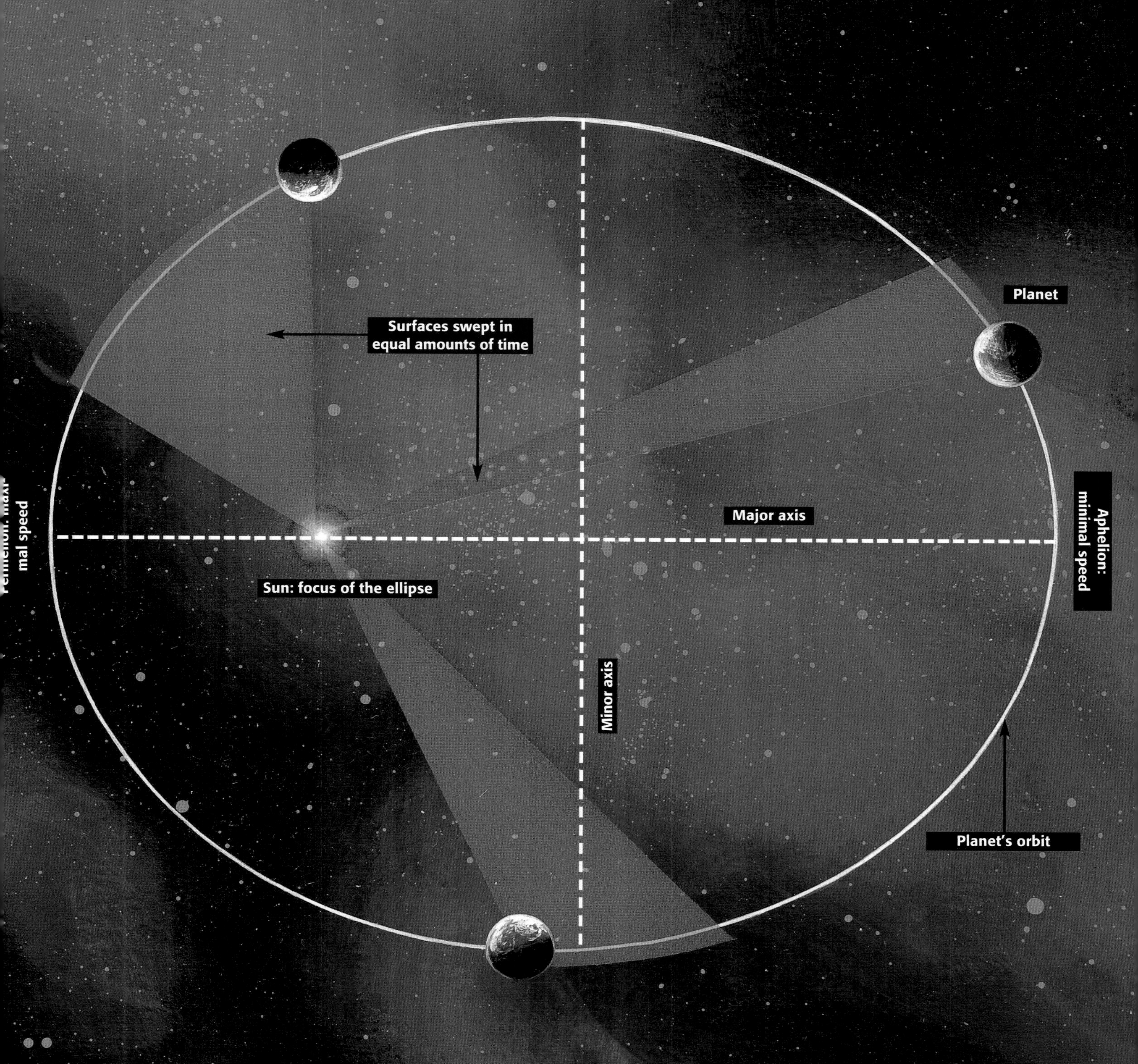

Planet
Surfaces swept in equal amounts of time
Aphelion: minimal speed
Major axis
Sun: focus of the ellipse
Minor axis
Planet's orbit

Kepler, his sorrowful wife Barbara at his side, joined his new teacher in Prague, now imperial mathematician to Emperor Rudolf II. Mars, however, refused categorically to submit to the tyranny of the many epicycles and eccentrics that peppered its theoretical orbit. Its various positions, minutely observed by Brahe, never matched his predictions. Kepler couldn't have come at a better time: whoever succeeded in determining the true orbit of Mars would hold the key to the mystery of the cosmos. But the prince of observers was a miser who released his data only with an eyedropper, the way he'd throw an occasional bone to his dwarf Jepp. Kepler insulted him, lost his temper, and returned to Graz, only to come back again a few months later. The two monsters lived together as well as could be expected, until Brahe's death in 1601. On his deathbed, the Dane begged Kepler to continue his work, so he wouldn't die in vain.

KEPLER'S LAWS

Johannes Kepler was named imperial mathematician almost immediately after Tycho Brahe's death. He seized his master's infamous measurements and started working relentlessly. Eight years later, in 1609, he published the *New Astronomy,* in which he lay down two of the three laws which ever since bear his name (see sidebar, pp. 114–115). According to the first, Mars's orbit—and by extension, those of all the other planets—is not shaped like a circle, but an ellipse. The second law stipulates that the speed of the planets changes as a function of their distance to the sun. The greater that distance is, the more the planet slows down, and vice versa. The mystery of the inequality of the seasons is now revealed: the earth is closest to the star in winter and furthest away in summer.

THE PHASES OF VENUS AND MERCURY

Mercury and Venus, located between the sun and the earth, are so-called inferior planets. They revolve around the sun in the same direction as the earth, counterclockwise. The first, the one closer to the sun, comes full circle—its sidereal period—in 87.96 days. The second takes 224.7 days. Both display a succession of phases resembling those of the moon. Venus is much easier to observe than Mercury, which is too close to the sun and therefore submerged in its light.

When their journeys lead them to move between the sun and the earth, they are said to be in inferior conjunction. They are at their closest point to us, at their maximal size and are invisible, offering us only their dark side. They continue their route and pass the earth, which is moving more slowly than they are. Venus begins to show herself in the morning at dawn, in the form of a thin crescent that fills itself out as she advances. Mercury and Venus hold the position of so-called greatest western elongation when they form a 90-degree angle with the earth and sun. Venus becomes half visible in the morning hours preceding dawn. She is entirely lit at the moment of superior conjunction, when she is exactly opposite the earth in relation to the sun. Venus can only be observed several days later, however, at the moment of sunset. She is furthest away from us and appears minuscule. Her lit portion is reduced to half when she reaches the position of greatest eastern elongation forming once again, a 90-degree angle with the sun and earth. The circle is completed when she returns to inferior conjunction. Careful, the amount of time separating the two successive superior conjunctions of Venus or Mercury is not equal to the sidereal year: it measures around 116 days in Mercury's case and 584 days in that of Venus. The earth, you see, is moving, too: when the planets return to where their previous conjunction took place, the earth is no longer there and the two planets have to catch up with it.

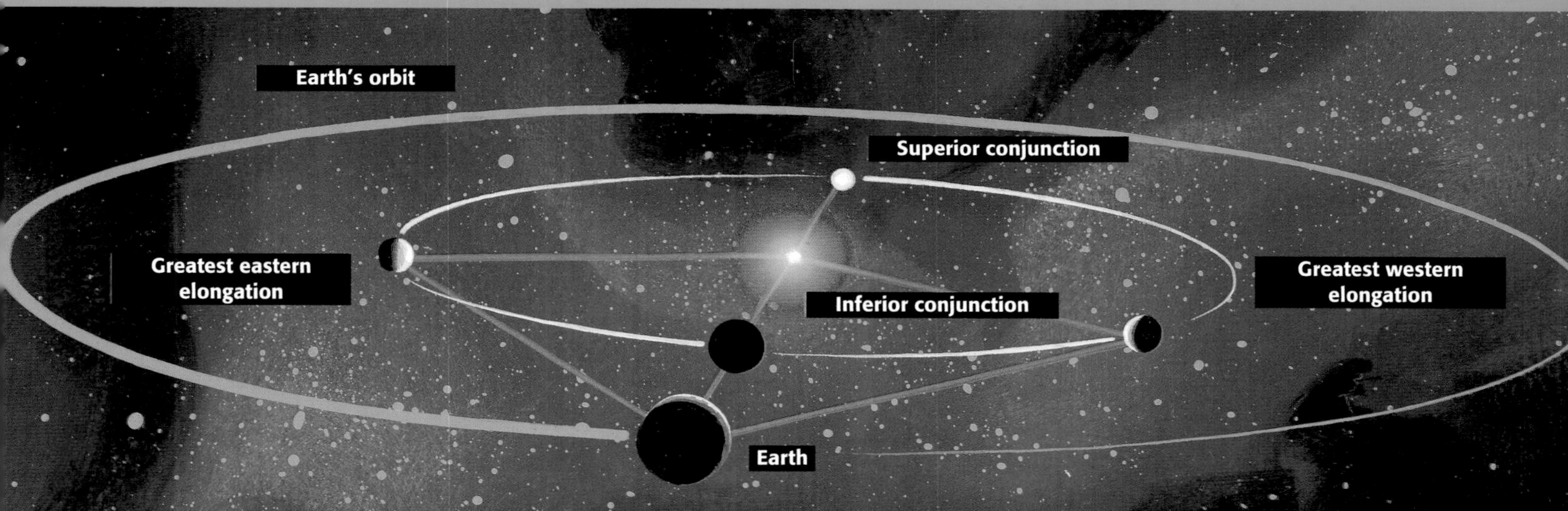

It's thus faster during the harsh season, which doesn't last as long as the fair-weather days. Appearances have been kept, Plato's three restrictions have been removed, and Ptolemy's system has been scrapped.

Galileo's Telescope

In 1609, Galileo, an Italian engineer and mathematician from Padua, decided to aim at the sky a telescope which he had personally improved. He discovered the four major satellites of Jupiter, contemplated the pockmarked surface of the moon and was buried beneath the avalanche of stars that form the Milky Way. In 1610, he tore a page out of the Copernicans' book when he discovered that Venus had phases (see sidebar, p. 117): this is impossible under the Ptolemaic system, where the planet would appear only in crescent shape. Kepler begged Galileo, who turned a deaf ear, to lend him a telescope (rather ironic, for it was Kepler who explained how lenses worked). The myope finally succeeded in procuring himself the instrument, and at last was able to gaze upon the heavens to which he'd dedicated his life.

Harmony Restored

Kepler's eldest son died of smallpox a year later. Then Rudolf II, in his turn, died leaving the astronomer without a protector. War was raging between Protestants and Catholics across Europe. Kepler moved to Linz in 1612. His wife plunged into madness and died. Yet Kepler still wouldn't let go, as determined as ever to discover the great secret of the cosmos: according to his third and final law, there effectively exists a mathematical link between the speed of the planets and the size of their orbits (see sidebar, pp. 114–115). The planets cannot change trajectory, under penalty of breaking all

equilibrium of what was from then on called the solar system. In 1618, Johannes Kepler published *The Harmony of the World* and exulted: "I have stolen the golden vase from the Egyptians and made it a tabernacle to my God [. . .]. I have thrown the dice in writing a book for either my contemporaries or for posterity. I do not care. A reader cannot wait a hundred years. God waited six thousand years for a witness."

The End of a Great Man

Kepler's life became hell. Misery moved into his new home, the emperor, for whom he was still mathematician, wasn't paying him. Kepler saved his mother from the stake at the last minute, where she had been led because of her over-familiarity with plants. He survived by making horoscopes on behalf of a duke, until, pushed to the limit, he left for Ratisbon to reclaim the twelve thousand florins the emperor owed him. He died there on 15 November 1530. Johannes Kepler, the man who'd truly changed the face of the universe, ended where he began, and was buried in a common grave.

Mercury, Venus, Earth, Mars, Jupiter, Saturn, Uranus, Neptune and Pluto . . . Nine planets revolve around the sun. In the same, counterclockwise direction. The planes of their orbits deviate very little from that of the earth. Pluto and Mercury have the most tilted trajectories, at seventeen and seven degrees respectively in relation to the ecliptic. Their orbits are also the most eccentric: Pluto's distance from the sun varies between approximately 2.7 billion and 4.6 billion miles, while that of Mercury varies between 28.6 and 43.5 million miles. Pluto's trajectory regularly brings it inside the orbit of Neptune.

THE SOLAR SYSTEM

All the bodies of the solar system revolve around their own axis, in the same direction as the earth, except Venus, which spins in the opposite direction. Uranus's axis of rotation is practically lying down: the planet spins like a wheel.

Venus and Mercury don't have satellites. Two small rocks, Phobos and Deimos, gravitate around Mars. The former gets inexorably close to the red planet, and will collide with it in a hundred million years. Jupiter is adorned with a necklace of over twenty satellites, including the very volcanic Io: wracked by massive tides, it is in a state of permanent eruption. Among Saturn's companions, surrounded by its magnificent rings of ice and rocks, Titan is the object of all us earthlings' attention because of its atmosphere. Uranus has eighteen courtiers, and Neptune has eight. The orbit period of Charon, Pluto's only satellite, is equal to the planet's rotation around its own axis: both continually look each other in the eye.

7

Celestial Mechanics

1. GRAVITATIONAL THEORY
2. CELESTIAL MECHANICS

Johannes Kepler's universe is divinely elegant. Freed from their circular shackles, now drunk with their new liberty, the planets glide gracefully around the sun. Kepler was far too shrewd not to wonder about the obscure link between our star and its satellites. Why do they all revolve around it? Why do they travel in ellipses? Why such submissive and adoring behavior? The sun is both the mathematical center of their movement and its underlying cause. It exerts some sort of force, a magnetic one, Kepler thought, which keeps the planets in their orbit. French philosopher René Descartes, too, had an inkling, albeit far less poetic, of the physical basis for the motion of celestial bodies. He believed that the universe is infinite, and filled with a rarified substance, a very subtle "ether." The stars and sun—which are made up of the same material—stir this substance up in their motion. The ether starts to swirl, and its eddies pull the planets along like corks upon water.

This vision of the universe, however, was torn to shreds by a young, reclusive, gifted and thoroughly incomprehensible English scientist, Isaac Newton. Standing on the shoulders of the giants who preceded him—Descartes, Kepler and Galileo—Newton replaced their world views with a universe in which every star, whatever its consistency, is subject to the authority of a singular, powerful force: gravity. This force, which dictates the tiniest movements of everything in the universe, allowed astronomers within its iron grip to finally get beyond merely describing celestial motion. They could now tap the source behind such movement directly, and forge a new and fearsomely effective tool for exploring the cosmos: celestial mechanics.

1.

RAVITATIONAL THEORY

Isaac Newton was born on 25 December 1642 (under the Julian calendar, that is, which was still in effect in England) in Grantham, a small city in Lincolnshire, England. He was so underdeveloped that his family, convinced he wouldn't survive, rushed to have him baptized. His father, a fat and illiterate landowner, had died three months before his birth; his mother, Hannah Ayscough, remarried a certain Pastor Barnabas Smith three years later. Little Isaac's new stepfather wanted nothing to do with him, and shipped him off to live with his grandmother. (Newton held a grudge against Smith his entire life, at one point threatening to burn him down along with his house.) Serious, withdrawn and

Young Isaac Newton beneath his apple tree. According to a legend of his own making, the genius discovered gravity at the age of twenty-three when he saw an apple fall to earth. In reality, he'd been reflecting on the subject for a good twenty years.

[© D. R.]

morose, Newton was no little angel. Very skilled with his hands, he built sundials and clocks, and enjoyed scaring the peasants half to death by launching weird, flaming missiles into the air.

Newton was nineteen when his mother sent him to Trinity College in Cambridge. He was a veritable glutton for knowledge: optics, theology, philosophy, mathematics. He touched on every subject, or almost every: the fine arts—sculpture, painting or music—left him completely cold. He immersed himself to such a degree in his personal research that he wouldn't hesitate to stick a piece of wood behind his eye in order to study the effects of retina distortion. At one point he had to lie prone for several days in the dark after almost losing his sight staring at the sun. Naturally, he didn't drink and never went out. In over five years his schoolfellows heard him laugh only once, when a student asked him what he was planning to do with a book by Euclid he was carrying.

NEWTON'S APPLE

In 1665, an outbreak of the plague emptied the college, forcing Newton to return home. Later on, when he was old and famous, he claimed that the idea of a universal force governing the motions of the moon and planets came to him that year, when he watched an apple fall from a tree.

The Inverse Square of the Distance

Galileo had beaten a clear path for Newton. The Italian had dedicated a major part of his work to celestial

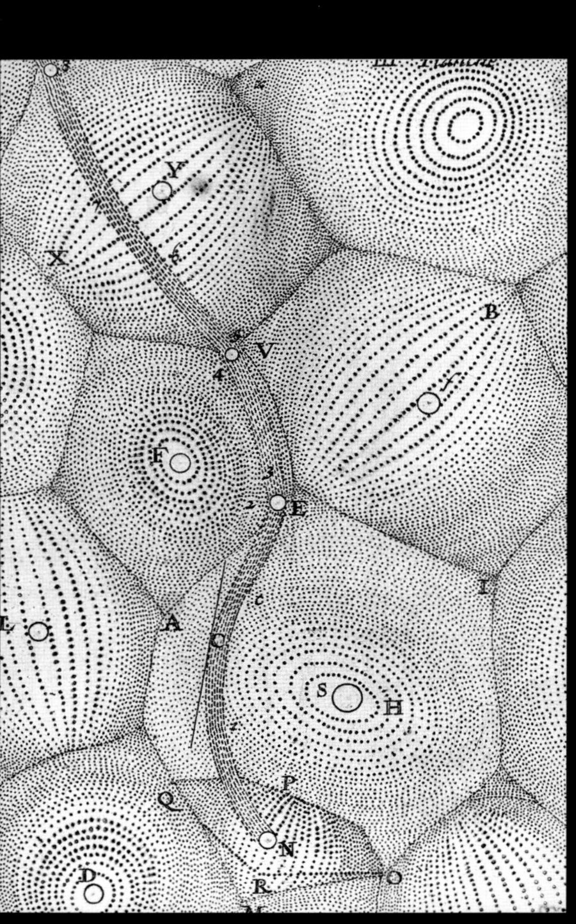

René Descartes's whirlpools. According to Descartes, vacuums in space are impossible, and there most be some sort of contact—an impact, a push—for motion to occur. The "substance" that fills space is rife with eddies

it down, stop it or alter its course. On earth, a marble rolled along a floor eventually stops because of friction. If we eliminate friction, it will keep on rolling indefinitely, without the need of outside help of any kind. Descartes clarified matters by laying down the so-called law of inertia, one of the pillars of classical physics: a body not subject to any force, any "impulse" likely to modify its speed and direction, is either at rest or moving with uniform rectilinear motion. The moon's natural tendency is to move not in a circle but in a straight line, Galileo believed, and Descartes invented the convenient but incorrect notion of ethereal whirlpools as a means of explaining its curved trajectory. Newton argued that the moon is subject to a second movement which bends its route, steers it towards earth and keeps it in orbit. Basing his argument on Kepler's third law, he asserted that this force must be proportional to the square of the distance separating the moon and the earth.

It's at this moment an apple broke away from its branch and hit the ground with an enormous cry. Newton had the revelation. Released at a given height, all solid bodies will fall to earth as if they were irresistibly attracted to it. What would happen if the apple fell from much higher up, say, from the height of the moon? Why isn't the moon, too, subject to the same force that causes apples to drop? Newton ruminated on this question, performed a few minor calculations, and arrived at the following conclusion: the force that pulls an apple to the earth but keeps the moon in its orbit is one and the same thing, and proportional to the inverse square of the distance separating either body from our planet. Eureka!

Philosopher and scientist René Descartes (1596–1650) sought a method for basing knowledge on certainties, for building a scientific edifice that would allow us to "master and possess nature." Newton was an attentive reader of Descartes's work, in particular mathematics.

[© Tours, Bibliothèque municipale/ *Ciel & Espace*]

"An Ellipse"

Newton's apple most certainly never existed. At the time, young Isaac was still far from formulating the universal law of gravitation. In a sudden bout of candor, he later admitted coming up with his master idea after "thinking about it all the time." He returned to Cambridge, where he was named professor in 1669.

Newton would work until he collapsed with fatigue, eat and sleep only when he remembered to, and would pace back and forth in his garden mumbling, his wig on backward. He'd teach before empty classrooms, but wouldn't notice. He would leave his numerous physical and mathematical discoveries (such as differential calculus) rotting away in desk drawers, would gladly answer questions provided he posed them, and never sought the recognition of his peers. A few colleagues did catch wind of some of his work and, in 1671, urged him to present before the Royal Society an invention of his destined for a promising future, the reflecting telescope. Newton took advantage of the occasion to defend his theory of colors, demonstrating that white light is not homogenous but formed from rays of different colors. The few criticisms he did receive provoked an attack of acute paranoia, causing him to withdraw again to Cambridge. The idea of a varying force like the inverse square of distance was in vogue, however. In 1679, Robert Hooke, a brilliant dabbler and member of the Royal Society who worked as a sort of headhunter, wrote Newton to ask him what the trajectory of a body subject to such a force might look like. In August 1684 astronomer Edmund Halley, who had already contacted Newton regarding the appearance of two comets in 1680, visited the hermit and posed the same question to him again. Newton spat out the reply: "An ellipse."

VERSAL GRAVITY

Three months later, Newton sent a small manuscript to Halley. In it, he demonstrated that the orbit traced by a body subject to a variant force, such as the inverse square of distance imposed by a second body could, in fact, be shaped like an ellipse. Ever so conveniently, Newton discovered that the latter respects Kepler's second and third laws perfectly. Halley begged him to publish these findings and Newton accepted, requesting a bit more time to review two or three minor details.

Force and Matter

The motion of any object can be characterized by its acceleration, that is, the direction in which it's moving and its speed. Force is the impulse necessary to transmit a given motion to a body, that is, an acceleration. To do so, this impulse must conquer the object's inertia, its natural tendency to remain at rest or in uniform rectilinear motion. For Newton, this inertia depends upon the object's mass, or the amount of matter it contains. From this he deduced the second great law of dynamics, according to which the intensity of a force is equal to an object's mass multiplied by its acceleration. This applies to gravity, which becomes proportional not only as a function of distance, but of the mass of the object attracted as well: at an equal distance, an object one hundred times more massive than an apple will be attracted by the earth with one hundred times more force.

Mutual Attraction

Now, according to a third great law articulated by Newton called that of action and reaction, an object subject to a given force by another body will return that force with equal strength. The apple, however puny it

[50]

SECT. III.

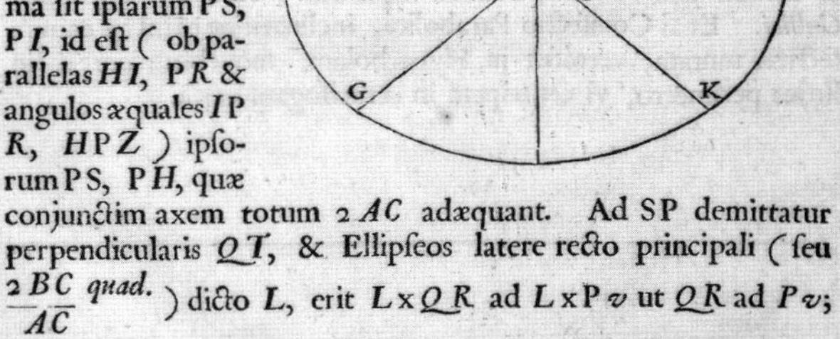

De motu Corporum in Conicis Sectionibus excentricis.

Prop. XI. Prob. VI.

Revolvatur corpus in Ellipſi: Requiritur lex vis centripetæ tendentis ad umbilicum Ellipſeos.

Eſto Ellipſeos ſuperioris umbilicus S. Agatur *SP* ſecans Ellipſeos tum diametrum *DK* in *E*, tum ordinatim applicatam *Qv* in *x*, & compleatur parallelogrammum *Qx PR*. Patet *EP* æ-

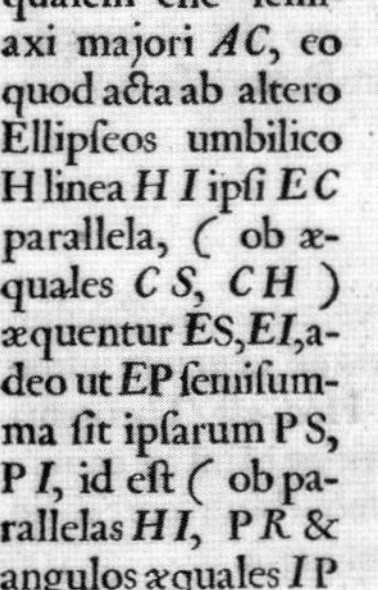

qualem eſſe ſemiaxi majori *AC*, eo quod acta ab altero Ellipſeos umbilico H linea *HI* ipſi *EC* parallela, (ob æquales *CS*, *CH*) æquentur *ES*, *EI*, adeo ut *EP* ſemiſumma ſit ipſarum PS, P*I*, id eſt (ob parallelas *HI*, P*R* & angulos æquales *I*P*R*, *H*P*Z*) ipſorum PS, P*H*, quæ

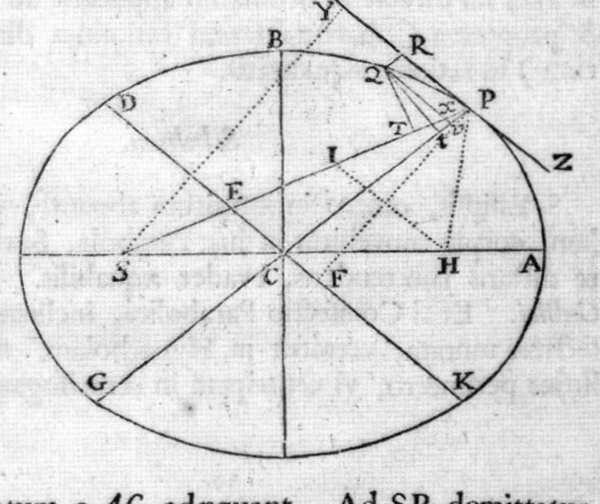

conjunctim axem totum 2 *AC* adæquant. Ad SP demittatur perpendicularis *QT*, & Ellipſeos latere recto principali (ſeu $\frac{2\,BC\ quad.}{AC}$) dicto *L*, erit *L*x*QR* ad *L*xP*v* ut *QR* ad P*v*; id eſt ut P*E* (ſeu *AC*) ad P*C*: & *L*xP*v* ad *Gv*P ut *L* ad *Gv*; &

Excerpt from Newton's *Principia*. Dry and filled with geometrical demonstrations, this work had an enormous impact.

[Isaac Newton, *Philosophiae naturalis principia mathematica*, 1687, Paris, BNF. © BNF]

may be, exerts its own force of attraction on the earth in the opposite direction and of an intensity equal to the one it endures. It's not a question of attractors vs. the attracted: the earth's attraction to another object is proportional to the two bodies' relative masses. The earth, however, won't begin spinning around the apple. It's far too big in comparison to the tiny fruit, which is perfectly incapable of setting it in motion. The sun, which itself contains more than ninety-nine percent of the total mass of all the objects in the solar system, is far too big for Venus's, Jupiter's or Pluto's attraction to have major repercussions on its movement. Finally, the force that causes apples and moons to tumble toward the center of the earth (and not toward its surface)—gravity, that is—is universal. It depends solely on the masses of the objects in question and not their substance. Gravity, therefore, is present everywhere you find mass in the universe. *In sum, all bodies attract each other with a force proportional to the product of their mass and the inverse square of the distance separating them.*

AND THERE WAS LIGHT . . .

Gravity explained everything, or almost: the elliptical trajectories of large planets like Jupiter and Saturn result from the enormous sun's force of attraction. This is the same gravitational net, moreover, that ensnares the moon. The stars can't escape this rule either: it's common for two or more of these heavenly bodies to begin circling around each other in a mutually attractive grip. Nothing is believed to ignore this law, not even galaxies, those gigantic blocs containing billions of stars that are worked and kneaded by forces way beyond the collective imagination of all the apples on earth. Newton had discovered the singular cause behind all celestial motion.

The Precession of the Equinoxes

Things might have stopped there, but Newton's gravitational theory lent itself as well to making a number of startling predictions. Under the combined effect of gravity and its rotation around its axis, the earth should be slightly flattened at the poles and somewhat thicker around the equator. In 1736, the French Academy of Sciences sent a die-hard "Newtonian," the mathematician Pierre-Louis Moreau de Maupertuis, to Lapland, in Finland. If the earth is flatter at the poles, then the length of an arc with an underlying angle of one degree should be smaller measured on the same meridian in Lapland than in France. And this is well the case: the earth is not an orange, but a grapefruit. Which, it seems, solved one of astronomy's deepest mysteries: the

The 1736 French expedition to Lapland measured the flattening of the earth. A previous team had been sent to Peru in 1735. One of its members, naturalist Charles Marie de la Condamine, didn't return until 1751!
[© Collection Viollet]

precession of the equinoxes (see chapter 3). Our planet is subject to the dual attraction of the sun and the moon, our unforgettable satellite. This effect is most noticeable at the level of the equatorial bulge, pulled upon as if sun and moon were trying to yank it back up to the ecliptic. Their efforts are in vain, however: rather than reduce the tilt of the earth's axis, they succeed only in inducing the axis to wobble like the stem of a top at the end of its spin. In the mid eighteenth century, English astronomer James Bradley discovered how the precession is complicated by the movement of the moon. While making its cone-shaped journey around the vertical pole of the ecliptic, the earth's axis makes tiny little circles strangely reminiscent of Ptolemy's epicycles. This is called "nutation."

Halley's Comet

Gravity met with its first great media success thanks to a comet. Comets are big balls of ice mixed with dust. They, too, revolve around the sun, following very elliptical trajectories. Their orbits regularly bring them very close to the sun, which causes their temperature to rise considerably. They then emit streams of gas and dust, forming their famous tails which can reach millions of miles long. In Newton's era, the movement of these bodies was a real mystery. In November 1680, the curmudgeonly genius rushed off to observe a huge and very bright comet that could be seen in the sky over the course of several days. A month later, a second comet streaked across the firmament, but in the opposite direction of the first. For Newton, armed to the teeth with his all-purpose theory, the two bodies were actually one and the same: the comet had merely made a hairpin turn behind the sun. A comet is subject to the same law as the planets; its orbit is merely much more elongated, with the sun located very near one of its extremities. Basing his argument on gravitational theory, Edmund Halley predicted that this comet should return in either late 1758 or early 1759. In 1757, French mathematician Alexis-Claude Clairaut refined Halley's calculations and called for its return in mid-April 1758, give or take a month. The comet, the famous one named after Halley, made its rendezvous in mid-March.

The Tides

The tides are one of the most spectacular consequences—from an earthling's point of view—of the gravitational marriage of the earth and moon. The force with which the satellite attracts our planet towards itself generates a wave, which in turn causes the surface of the oceans to fluctuate between a maximum and minimum level. The sun's influence lets itself be felt most strongly at the new and full moons, when it sits in alignment with the satellite. The tides, more pronounced at these moments, are called spring tides. During the first and last quarters, conversely, both bodies yank the earth in different directions and the tides, called neap tides, are not as strong. The most powerful tides take place on the equinoxes, when the sun is directly vertical to the equator. This phenomenon also affects the solid part of the terrestrial globe, which "inflates" and "deflates" to the same rhythm as the oceans. These deformations are not elastic, however: the earth, due to the constant rubbing of the water against the ocean floor and frictions between its various internal strata, is slowly losing energy. Its rotation is affected: little by little, our planet is slowing down. Approximately 400 million years ago, the day lasted

Famous Halley's Comet has been observed many times since antiquity. Its trajectory was established in 1705 by astronomer Edmund Halley, Newton's faithful first supporter. It comes back to visit us on average every seventy-six years.

[© *Ciel & Espace*/A. Fujii]

only twenty-one hours. This braking motion has had rather strange repercussions on the moon as well: our satellite is drifting further away from us at the mind-boggling speed of 11.5 feet per century.

SIR ISAAC NEWTON

Isaac Newton's first great work appeared in 1687—on Halley's dime—under the title *Mathematical Principles of Natural Philosophy*. Its success was almost immediate, thanks to faithful Halley's aggressive propaganda. The book was not an easy read, however, and Newton, who held most mere mortals in utter contempt, claimed he had intentionally overcomplicated his demonstrations to make his theory accessible only to a handful of experts. He perhaps did so, as well, as a slight to Robert Hooke, who, though hardly partial to mathematics, had also claimed to have discovered gravity. Such a claim outraged the irascible Isaac, who had nearly avoided publishing anything at all; Newton made himself the bane of Hooke's existence until the latter's death in 1703. The genius could prove to be as petty as he was cruel: appointed warden of the Royal Mint, Newton sent to the galleys and gallows every counterfeiter he could get his hands on. Ennobled in 1703, then president of the Royal Society and later a member of Parliament, the former hermit polished up his image, surrounded himself with a court of admirers, offered portraits of himself to his guests and directed all his bile towards poor Gottfried Wilhelm Leibniz. The latter, a great mathematician of German origin, had made the mistake of publishing a method for calculating the infinitesimal without giving credit to Newton. This quarrel, too, only ended with Leibniz's death in 1716. Sir Isaac Newton loved no one but his mother, the only person who had access to his rare displays of affection. He died a virgin in 1727.

2.

CELESTIAL MECHANICS

Universal gravitation is, in theory, a formidable tool for astronomers. Rather than estimate the positions of planets based on their supposed trajectories alone, astronomers now needed only to calculate the various gravitational influences a planet might be subject to and analyze their impact on its movement. Starting from the initial cause directly, the study of planetary movement could only gain in precision. In practice, however, it took nearly a century for this new approach to astronomy, celestial mechanics, to become operational.

ALL BODIES ATTRACT . . .

Newton opened a veritable Pandora's box. Gravity is sublimely simple when you're only dealing with two bodies, especially when one of them is very large, like the sun, and the other fairly small, like a planet. Subject to powerful solar attraction, planets quietly revolve around the star, following paths that respect Kepler's three commandments to the letter. Of course, they exert in return a small retaliatory effect on the sun, but this isn't too harsh and is quite easily calculated. Up to this point, everything's fine. It turns out that the solar system consists of nine planets, among which certain ones, such as Jupiter, have a relatively looming presence. In accordance with Newton's law, the planets all affect each other's movement. The earth, subject to the strict solar regime, must endure the gravitational exactions of Jupiter, Mars and the moon as well. Neptune jostles Uranus and Pluto, who return the gesture in kind. Io, Jupiter's satellite, simultaneously pulled in a tug-of-war by her planet and companion moons, has become nothing more than a volcanic scab.

Orbits Under the Influence

Under the effect of all these forces, the planetary orbits never cease changing shape and no longer resemble their Keplerian ideals. To know a planet's position at a given moment, you need to calculate all these gravitational actions and determine their consequences on its trajectory. The problem is further complicated by the fact that the members of the solar system never stop moving in relation to one another. The intensity and direction of a Jupiter or a Saturn's attractive force on the earth, for example, vary constantly. One of the worst cases is that of the moon: her orbit, thanks to the countless blows inflicted on her by the earth and sun, doesn't have a well-defined form. She's constantly swelling, shrinking, fluctuating and undulating (see sidebar, pp. 137–139).

THE THEORY OF DISTURBANCES

Grasping the size of the task, Sir Isaac personally threw in the towel. A small tribe of gifted mathematicians—the Swiss Léonard Euler, the Frenchmen Jean Le Rond d'Alembert, Alexis-Claude Clairaut, Joseph Louis Lagrange and Pierre Simon de Laplace—took up the challenge. Each added a brick to the monumental edifice of celestial mechanics, which was nearly complete by the late nineteenth century. The sun is the largest of attractors and its action is predominant. All by itself, it would have imposed a perfectly Keplerian trajectory on the planets. The parameters of this ideal orbit—eccentricity, major and minor axes, tilt—are all fairly easy to calculate. You need then consider how this orbit is disturbed by the gravitational effects of the other celestial bodies: the object of the game is thus to isolate and to identify, if not all, then at least the most significant of these disturbances, and to account for them in the calculation of a given planet's movement (via so-called Lagrange equations).

Orbits of Varying Shape

These disturbances make themselves known through changes in the parameters of the theoretical ellipse. An orbit's eccentricity, that is, its shape, alters over the course of time, as well as the orientation of its major axis and its tilt in relation to the ecliptic. These alterations are periodic; they aren't permanent changes but temporary. The eccentricity of an orbit increases very slowly, reaches a certain maximum value and then begins to decrease. The ellipse gets rounder, then the process reverses itself and it starts flattening out again. Similarly, the orientation of the major axis changes in such a way that it always eventually returns to a position it held some time before. The duration of these cycles can vary greatly and depends on the disturbance in question: certain disturbances dissipate after several hours or days, while others may last for tens of thousands of years.

The Discovery of Neptune

On the night of 13 March 1781, musician, amateur astronomer and master telescope manufacturer William

A planetary mobile. This delicate object, useful only as a teaching tool or decoration, is far more informative than the abstract equations of celestial mechanics, as it provides a visual reenactment of the planets' motion around the sun.
[Paris, musée du CNAM. © *Ciel & Espace*/ E. Perrin]

Herschel was navigating in the sky as was his custom. He observed a hazy little object in the constellation Gemini. He decided to track it, and soon noticed that the thing was moving in relation to the stars. Herschel, who thought it might be a comet, requested the help of professional astronomers, who congratulated him on his discovery: it was a planet. He named it Georgium Sidus in honor of King George III of England, who offered him a small pension. The continent, however, didn't care for this name at all and had it changed to Uranus. This planet's trajectory is terribly disturbed, much more than it should be were it truly the outer planet of our solar system, as everyone assumed at the time. Things ended there until mathematician Urbain Jean Joseph Le Verrier decided to look into the matter. Armed with a pencil and a sheet of paper, he worked out all the figures and soon declared that Uranus's troubles were ascribable to an eighth planet, whose position he provided in 1846. On 23 September of that year, Neptune was observed for the first time. John Couch Adams, a young student at Cambridge, arrived at the same conclusion as the Frenchman, but His Majesty's astronomers didn't bother to verify it. Le Verrier tried to repeat his feat with the planet Mercury, whose movement also presents certain anomalies. He concluded that Mercury was under the influence of another small planet, Vulcan, which set astronomers off on a wild-goose chase. The search for Vulcan was a vain one: the theory of general relativity easily explained Mercury's deviations in the early twentieth century. For his efforts, Le Verrier was named director of the Observatory of Paris in 1854, where his terror reigned for more than twenty years.

THE MOON

The Movement of the Moon

Our dear moon measures 2,160 miles in diameter with a mass equivalent to 0.0123 times that of the earth. It doesn't stand straight on its axis, which is at an angle of 83°19′ to the plane of the lunar orbit. The latter has an average tilt of 5°8′—which varies by 0.3° every 173 days—in relation to the ecliptic. The moon's distance fluctuates between 221,438 miles at the perigee, the point in its orbit when it's closest to the earth, and 252,724 miles at the apogee, the point when it's furthest away. The lunar orbit's average eccentricity is 0.0549. This is far from constant, however, and swings between 0.0666 and 0.0432 with a periodicity of 412 days. To exaggerate terribly, it's as if it regularly changed shape from that of a rugby ball to nearly that of a basketball. This instability varies based on the sun's position in relation to the moon. When, during the new moon, the moon slips between the sun and the earth, it naturally tends to get closer to the former and move further away from the latter. During the full moon, it takes a few steps back towards us. The solar attraction is even stronger when the moon is both new and located at the apogee. The major axis of its orbit points directly toward the sun and the flattening of its ellipse is maximal. This axis doesn't maintain the same orientation over time. It shifts and drifts, in the same direction as the moon, by 0.114° per day. It returns to a given position and points toward the same star only every eight years and 310 days. Thanks to all these disturbances, the moon's speed, which averages 635.66 miles per second, isn't constant. The moon is always late or early in relation to the theoretical time the new or full moon should take place.

The moon's orbit

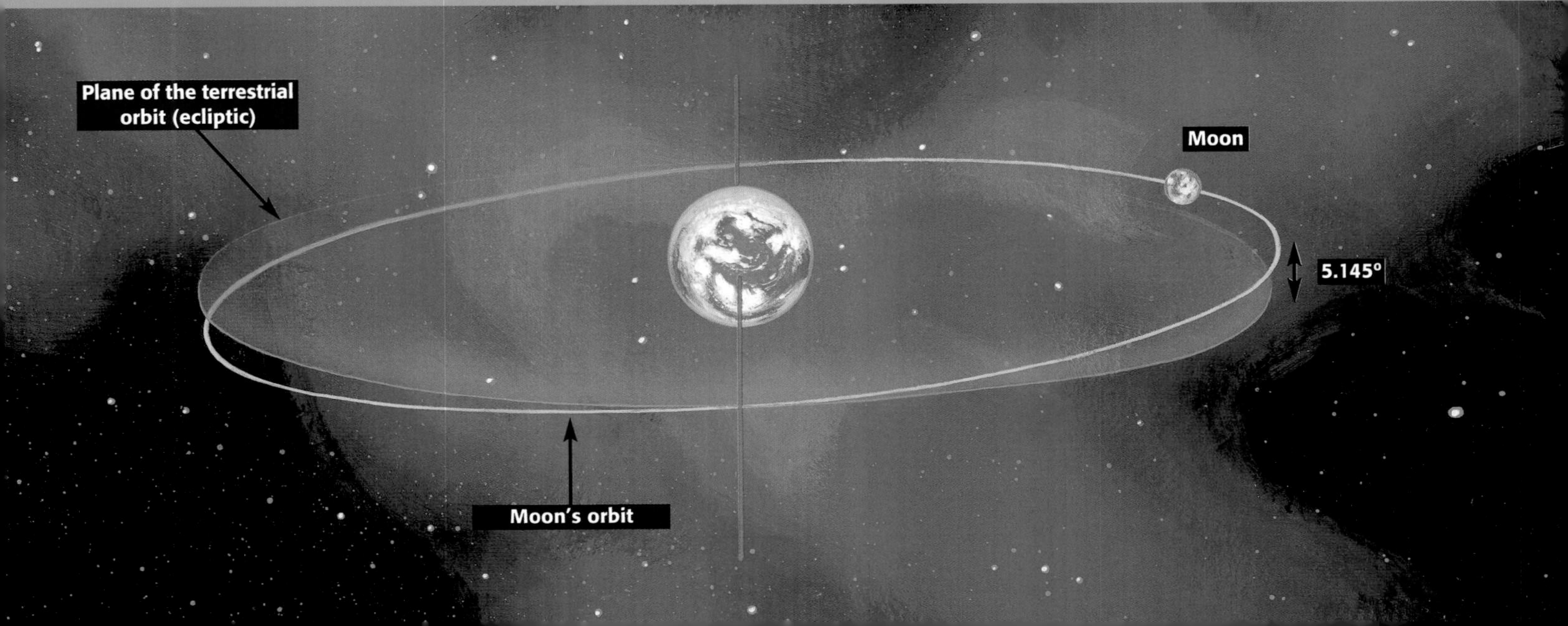

Plane of the terrestrial orbit (ecliptic)

Moon

5.145°

Moon's orbit

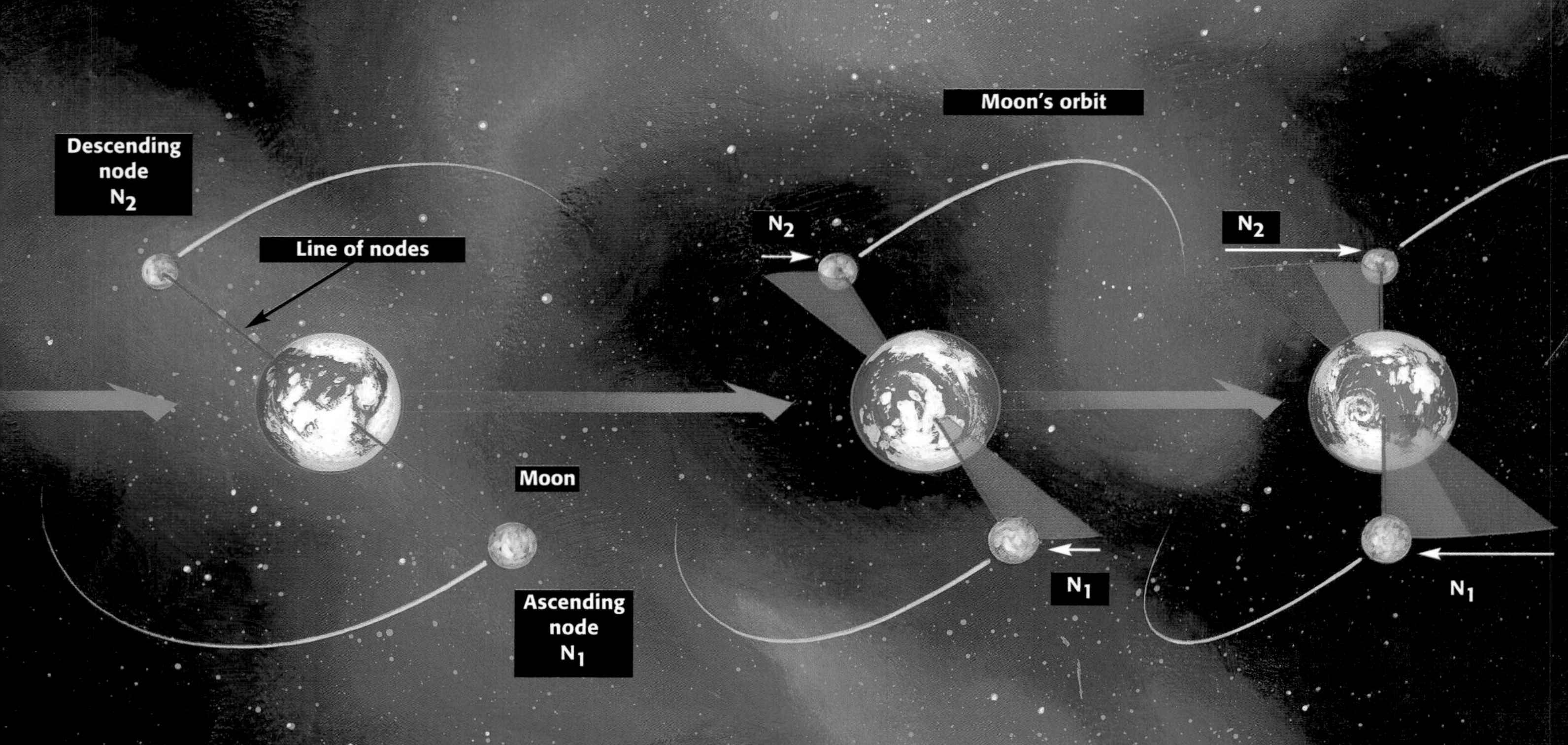

Rotation of the line of nodes

A Fistful of Nodes

The moon's orbit is tilted about 5 degrees in relation to the earth's ecliptic. The moon is usually above or below it, its trajectory crossing the plane at only two points, called nodes.

The nodes aren't fixed. Consider the moon at the moment it's climbing back toward the ecliptic. It crosses the plane at the so-called ascending node, climbs up to the highest point in its trajectory, then heads back down and passes through a second, so-called descending node. Common sense dictates it will return to its point of departure, the ascendant node, and that's what it does. But the latter has moved; it has stepped back, parallel to the ecliptic, about 1.5 degrees. Relative to its previous position, it has moved a distance equivalent to three times the moon's diameter. The descending node is affected by a similar movement and moves forward 1.5 degrees. Everything behaves as if, as time goes by, the nodes themselves were tracing a circle around the earth. They only return to their point of origin every 18.6 years. The result: the moon's trajectory isn't closed. If we were to draw it over the course of one rotation period of its nodes, or 18.6 years, we would end up with a veritable knot of orbits, a multitude of overlapping ellipses.

Predicting Eclipses

A solar eclipse occurs when the moon slips between the earth and the sun. The moon casts a cone-shaped shadow across our planet. When this shadow falls upon the earth's surface, the sun becomes totally invisible for everyone standing within the affected region.

An eclipse is possible only under several conditions. First, it can only occur when the moon is new. Next, for its shadow to fall squarely on the earth, the moon cannot be too high or too low in relation to the ecliptic. Similarly, it cannot sit either too far right or left of the line connecting the earth and sun. To accurately predict when a solar eclipse will occur, you need to figure out at precisely what moment all three conditions will be met. But that's not all: you must also specify where on earth it will be visible, that is, the terrestrial regions that will be touched by the narrow brush of the lunar shadow. This will vary greatly depending on whether the moon is located only a hair's breadth above or below the ecliptic. In short, the moon's position needs to be known with great precision. Because of the innumerable disturbances in its movement, eclipses aren't periodic. To the great annoyance of astronomers anxious for simple predictions, they don't reoccur at regular intervals.

The mechanism of eclipses

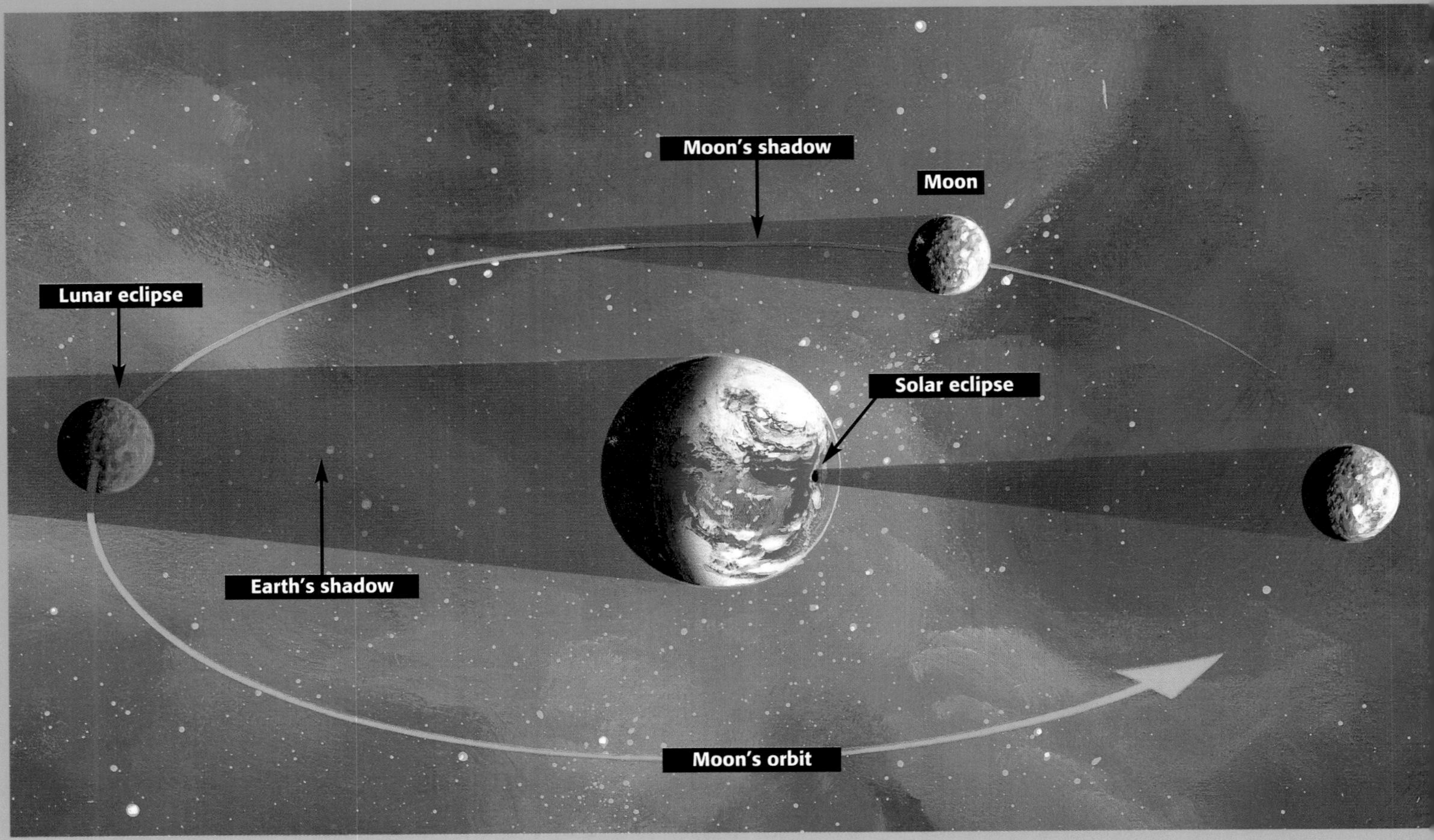

The planet Neptune was reportedly observed for the first time by Galileo, who mistook it for a mere star. Located around 2.8 billion miles from the sun, it's a gaseous giant seventeen times more massive than our planet. Its beautiful blue color is nothing like that of the ocean, but caused by the methane in its atmosphere.
[© *Ciel & Espace*/JPL]

Is Everything Predictable?

Celestial mechanics exulted: its methods proved so efficient that it became possible to study the planets based not only on the disturbances they endure, but those they inflict as well. The first extra-solar planets, discovered only a few years ago, could only be located thanks to the minute effects of their attractive force on the stars they circle.

The evolution of mathematics, the enormous precision with which today's instruments provide the positions used to determine planetary orbits, the power of computer calculations to account for ridiculously tiny disturbances—everything, today, reinforces the all-powerful image of celestial mechanics. But whatever may be the precision of its measurements, science will always have its limits. There will always exist some uncertainty, however slight, as to a heavenly body's exact position. According to chaos theory, the gap between reality and calculations will only get worse as time goes by. If an astronomer misses a measurement by thirty feet, a practically imperceptible mistake in astronomical terms, such an error can grow to millions of miles after several hundred million years.

THE THEORY OF GENERAL RELATIVITY

Isaac Newton always refused to answer a question that tormented physicists up to the early twentieth century: what causes gravity? What is this force that acts both at a distance yet instantaneously? Why does it exist? The credit for explaining what gravitation really is goes to Albert Einstein, as sociable, talkative and seductive a genius as Newton was misanthropic, silent and chaste. Einstein's 1915 theory of general relativity is far less accessible to the average person than the one of the illustrious Englishman. Space, according to Newton, where gravity is at work, resembles a sort of empty and infinitely-sized shoebox sprinkled with matter. Time, meanwhile, which merely gives rhythm to the existence of celestial bodies, is a separate matter entirely. According to Einstein's theory, in contrast, time and space are one and the same thing: space-time. With its four dimensions, space-time is perfectly inaccessible to our senses, which are used to moving about in only three dimensions. It's as if, for us, space and time were woven together into some unknown substance, a sort of mathematical cloth, which naturally distorts itself in the presence of matter. The enormous mass of the sun, for example, forms a large depression in the fabric of space. The planets travel through a space-time filled with dips and bumps that affect their trajectories. The planets act like balls quickly moving across an uneven surface: they sometimes roll around obstacles, or drop down into ditches and just as soon reemerge. The orbits of the nine members of the solar system are not the result of some special force, but governed by the morphology of space-time, twisted and distorted by the sun. Newton's conception of gravity, however, remains an excellent approximation of Einstein's gravity. Astronomers can always resort to it to determine, for example, the different orbits of artificial satellites, or to calculate the trajectories of interplanetary probes.

One day, a journalist reportedly asked astronomer Arthur Stanley Eddington, an ardent defender of the theory of relativity, if it was true that only three people in the world understood it. Eddington allegedly answered: "Who's the third?" Albert Einstein won the Nobel Prize in 1922, but for work other than on relativity.
[© IMS/AKG, Paris]

geocentric system of the universe.
[Bartolomeo Velho, *Principio de verdadeiro cosmographia et geographia universal de todas as terras que sao descubertas . . .*, 1568. Paris, BNF. © BNF]

The ancient universe was thick-skinned. The sphere of fixed stars withstood the serious blows dealt to Ptolemy's celestial geometry and Aristotle's clever cosmic scaffolding throughout the sixteenth and seventeenth centuries. The trajectory of comets, the moon's pock-marked face, Venus's phases and sunspots, however, eventually got the better of the concentric and material cocoon inside which the earth had been huddled. But humankind still wasn't ready to confront the void, and clung to its reassuring, starry lid for fear of tipping out into terrifying infinity. That is, until the fixed stars themselves came unhooked from their spherical wall and took off to cast their luminous train across the four corners of space. The stars regained their freedom of movement, but we observers didn't notice: transported by the silent turning of a phantom sphere, we continued to raise the same enamored eyes toward the sky as did the Cro-Magnons and Ptolemy.

1.

THE COSMIC ONION

The first to respond to Plato's call—"We must keep up appearances"—was his own student, Eudoxus of Cnidus. There is nothing truer to Platonic criteria than the movement of a sphere in smooth rotation around its axis. The trajectory of a planet solidly fixed upon such a sphere would appear perfectly circular and smooth at the center of the marble, occupied by the earth. Eudoxus imagined a model of the universe composed of a number of embedded, hollow spheres, with our planet at the center. Seven of these spheres carried the planets, sun and the moon, with the stars nailed to an eighth. The motion of the planets was governed by a clever combination of movements of several starless spheres, which influenced and pulled each other along mutually (see chapters 1 and 6).

ARISTOTLE'S WORLD

Nothing suggests that Eudoxus of Cnidus believed in the existence of his homocentric system. It's more likely that he used it as an artifice, a completely hypothetical aid destined to satisfy the very demanding Plato. This world designed like an onion, with its many spherical layers, took on greater meaning, body and consistency with Aristotle. Born in Stagira, in Macedonia, in 384 B.C., the future philosopher studied at the Academy founded by Plato and directed at the time by Eudoxus. Aristotle later became Alexander the Great's tutor from 343 to 340 B.C., before founding his own school, the Lyceum in Athens. The philosopher, who died in

of the moon, and the one beyond it, the superlunary. The former, which contained the earth, was composed of the four famous elements–air, earth, fire and water. This was the realm of corruption and changes of all kinds, where objects never ceased to age, rot, die and transform. It was quite the opposite for the heavenly bodies–the stars and planets–who were allergic to all modification: their appearance and movement had been the same since the dawn of time and everything suggested they were eternal.

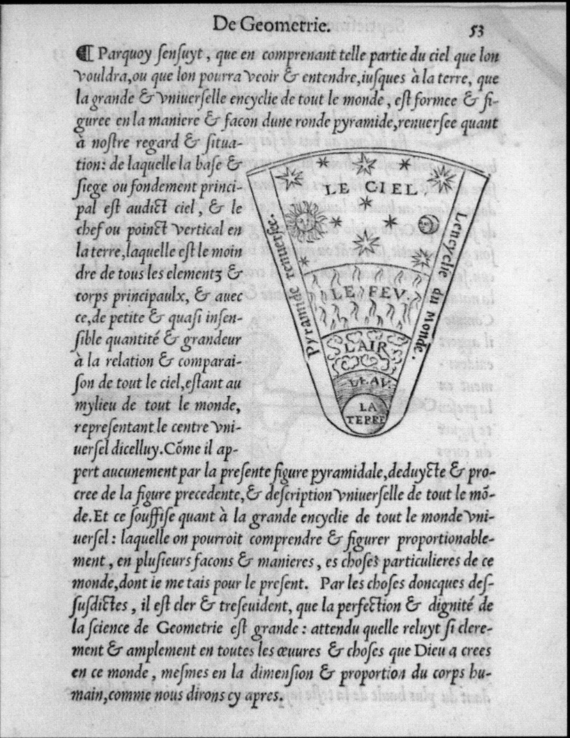

The Fifth Element

According to Aristotle, the movement of a body unaffected by any exterior force, any action likely to modify its trajectory, depended on its nature. In the sublunary world, air and fire had a natural tendency to rise, while earth and water tended downward, towards the earth. The heavenly bodies, in contrast, were uniquely partial to circular motion. More precisely, it was the spheres onto which they were fixed which spun round: Aristotle refused to accept the idea that celestial bodies could move about freely. The spheres, therefore, couldn't pos-

De Geometrie. 53

Parquoy ſenſuyt, que en comprenant telle partie du ciel que lon vouldra, ou que lon pourra veoir & entendre, iuſques à la terre, que la grande & vniuerſelle encyclie de tout le monde, eſt formee & figuree en la maniere & facon dune ronde pyramide, renuerſee quant à noſtre regard & ſituation: de laquelle la baſe & ſiege ou fondement principal eſt audict ciel, & le chef ou poinct vertical en la terre, laquelle eſt le moindre de tous les elementz & corps principaulx, & auec ce, de petite & quaſi inſenſible quantité & grandeur à la relation & comparaiſon de tout le ciel, eſtant au mylieu de tout le monde, repreſentant le centre vniuerſel dicelluy. Cõme il appert aucunement par la preſente figure pyramidale, deduycte & procree de la figure precedente, & deſcription vniuerſelle de tout le mõde. Et ce ſouffiſe quant à la grande encyclie de tout le monde vniuerſel: laquelle on pourroit comprendre & figurer proportionablement, en pluſieurs facons & manieres, es choſes particulieres de ce monde, dont ie me tais pour le preſent. Par les choſes doncques deſſuſdictes, il eſt cler & treſeuident, que la perfection & dignité de la ſcience de Geometrie eſt grande: attendu quelle reluyt ſi clerement & amplement en toutes les œuures & choſes que Dieu a crees en ce monde, meſmes en la dimenſion & proportion du corps humain, comme nous dirons cy apres.

the amount of space allotted to a given heavenly body including the set of spheres necessary to account for its movement. These skies were concentric and ran together, and there were as many of them as there were planets. It was absolutely impossible to jump from one to the other, with the sky housing the stars being the final rampart against nothingness. Aristotle abhorred the infinite almost as much as he did vacuums, and conceived the sphere of fixed stars as being of a reasonable, though indeterminate, size. The entire system was set in motion by some ethereal principle, a sort of driving intelligence stripped of corporal existence.

Ptolemy set aside the homocentric model, comprised uniquely of embedded spheres but unable to account for observation. The system was completely powerless to explain variations in the distances of the planets, for example. The Alexandrian astronomer ultimately endeavored to describe the trajectory of celestial bodies, stringing together epicycles and eccentrics without worrying much about their underlying cause. But Aristotle's authority was so great that the general principle of a spherical universe, with the earth at its center and replete with concentric and ethereal skies in which moved the heavenly bodies, was taken for granted.

The five elements according to Aristotle, ordered according to decreasing "heaviness"—earth, water, fire and air. The cosmos, from the moon to the sphere of fixed stars, was composed of a fifth essence, ether.

[Charles de Bouvelles (canon of Noyon), *Livre singulier et utile touchant l'art et practique de géométrie, composé nouvellement en françoys, par maistre Charles de Bouvelles . . .*, Paris, BNF. © BNF]

The Onion and the Holy Scriptures

In Johannes Kepler and Tycho Brahe's era, the Catholic Church wound up adopting this onion and its instruction manual, the philosophy and physics of Aristotle, as the official model of the universe. Some touching up

On one side, Aristotle and Ptolemy. On the other, the Bible . . . The theologians at the end of the Middle Ages tried to please everybody, by reorganizing the Greek sky in such a way that it would accept all those who, according to the Scriptures, resided there (saints, angels, the chosen ones . . .).

[Hartmann Schedel, *Das Buch der Croniken*, 1493. Paris, BNF. © BNF]

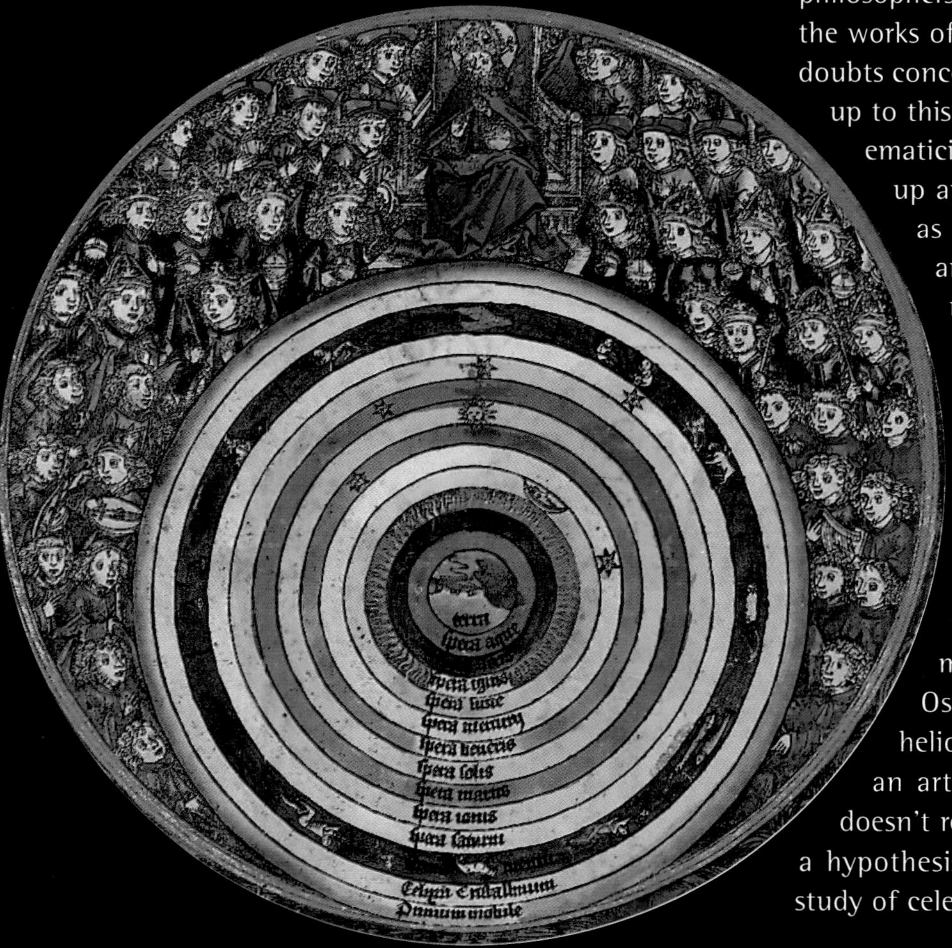

rendered it truer to the Holy Scriptures—a heaven, Paradise, took its place beyond the sphere of fixed stars and served as residence for the chosen few. It was preferable not to question those characteristics directly justified by biblical citations, such as the movement of the sun and the immobility of the earth.

Discussions and speculation surrounding the organization and veritable nature of the universe had become the private hunting grounds of theologians and philosophers, versed respectively in the Scriptures and the works of Aristotle. Some thinkers did, in fact, voice doubts concerning the validity of the postulates leading up to this image of the cosmos. On their end, mathematicians were uniquely responsible for keeping up appearances; they had the right to toy with as many circles and ellipses as they pleased and even let the earth move, provided they never presented their constructions as representing truth.

THE SPHERES IMPLODE

It's therefore hardly surprising that Nicolaus Copernicus, a Church canon no less, waited until he was at death's door before making his model public. His disciple Joachim Rheticus had entrusted the manuscript to a theologian friend, Andreas Osiander, who added a cautious preface: heliocentrism mustn't be taken literally, but as an artifice, a mathematician's trick. The earth doesn't really revolve around the sun; this is merely a hypothesis designed to simplify and facilitate the study of celestial movements.

Galileo's Manifesto

Italian physicist Galileo vigorously fought this way of thinking. There was no reason why the models elaborated by astronomers—who were scientists—should be mere games. Truth is inscribed in nature, not books, even if you're dealing with works by thinkers of Aristotle's caliber. A universe isn't constructed uniquely based on metaphysical considerations or general axioms, but on observation as well. Galileo applied this principle to studying the movement of the heavens and demonstrated how the latter obeys a few, simple laws. These laws are expressed in what Galileo called the language of nature: that is, mathematics.

The Skies Blend Together

The planetary spheres couldn't withstand the series of comets that appeared at the end of the sixteenth century. The very precise Tycho Brahe discovered that the one in 1577 leaped across the allegedly impassable skies, traveling effortlessly from one to the other. The comets behaved in total contradiction of Aristotle, in whose eyes they would have certainly counted among the sublunary riff-raff: in the superlunary world, things were never supposed to change. For Brahe and Kepler, this marked the end of the cosmic Russian nested dolls. In 1609, the telescope delivered another blow to Aristotle's world view, while providing evidence in support of Copernicus's. Galileo discovered that the moon is a horrid little gray world, as full of craters and bumps as the earth. It's nothing like the perfectly smooth and incorruptible marble described by the philosopher: later observation of sunspots confirmed that the distinction between the sublunary and superlunary worlds was no longer valid. Venus, who should have been limited to a

A striking portrait of Galileo by Italian artist Guido Reni. Although the scientist hadn't demonstrated the earth's rotation around its axis, he did demolish the proofs of the earth's immobility put forth by Aristotle and his partisans. In doing so, he provided the foundation for a science destined for a brilliant future, physics.
[© AKG, Paris]

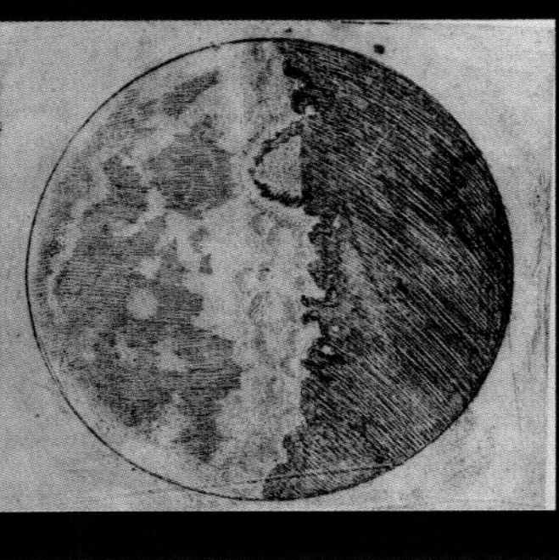
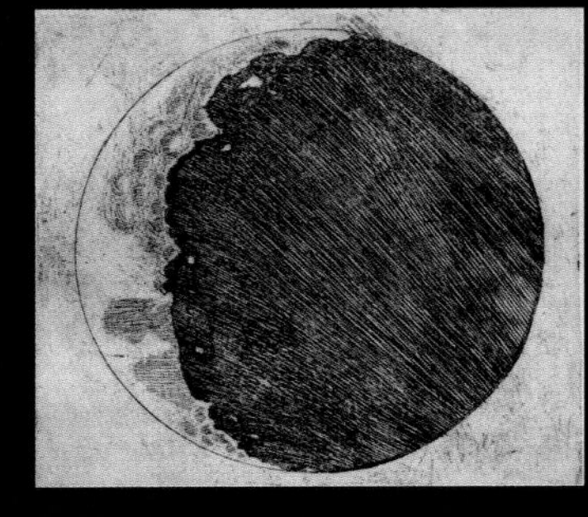
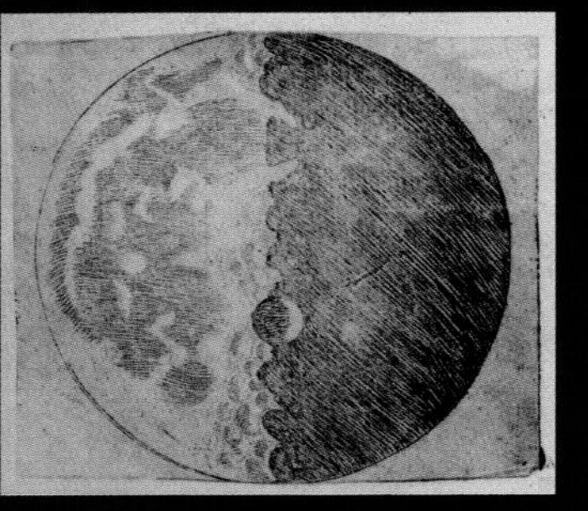
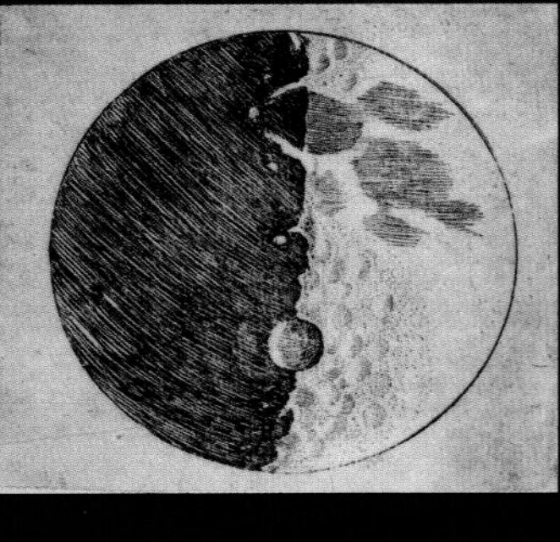

These drawings of the moon were made by Galileo in 1610. Quite a shock: This marked the first time people had seen the true face of our satellite. Galileo's observations were published in *The Starry Messenger,* a revolutionary work that earned him sudden, dangerous fame.
[Florence, Biblioteca nazionale. © Scala]

permanent crescent in Ptolemy's system, revealed herself to Galileo in all her phases. Shaken by these revelations, certain men of the cloth began hoping to adopt Tycho Brahe's system. This system was true enough to observations and had the distinction of leaving the dogma of the earth's immobility unchallenged.

Galileo's War

Becoming famous, Galileo entered into the service of the powerful grand duke Cosimo de' Medici. Emboldened by both the duke's protection and the kindness of an enlightened wing of the Vatican, Galileo waged a campaign to have heliocentrism considered as the truth, and not merely some mathematical artifice. Mocking and arrogant, he antagonized Aristotelians as well as part of the clergy by declaring that Copernicus and religion weren't mutually exclusive. In 1616, the Italian astronomer was reprimanded by the Church, which decreed that Copernicus's theory, in causing the sun to stop moving, was contrary to the Holy Scriptures. Friends begged Galileo to remain calm. He breathed easier with the accession of Pope Urban VIII, who had intervened in 1616 to prevent a total ban on Copernicus's book. In 1632, Galileo wrote *Dialogue Concerning the Two Chief World Systems*, where he compared the Ptolemaic and Copernican models. In it, Galileo used his own work on the earth's rotation to refute arguments advanced by Aristotle to justify the planet's immobility. In doing so, he derided partisans of the philosopher, whom he presented with the features of a simpleton—a caricature the Pope took to be of himself. Urban VIII had been weathering a terrible political and personal crisis, and needed to reaffirm his own authority and that of the

He therefore unleashed the Inquisition. Galileo, an old man of seventy, was summoned to Rome in 1633, and ordered to renounce, deny and reject heliocentrism wholesale. His *Dialogue* was forbidden along with Copernicus's book, and he was placed under house arrest at his home near Florence. Galileo went blind in 1638; he passed away in 1642. Before dying, he had time to publish *Discourses and Mathematical Demonstrations Concerning Two New Sciences*, the foundation for a new mechanics which would be taken to its heights by Newton.

Jupiter and its four major satellites—from left to right: Callisto, Ganymede, Io and Europa. They were discovered by Galileo who named them "Medicean Stars" in homage to his patron and protector, Duke Cosimo de' Medici.
[© *Ciel & Espace*/A. Fujii]

2.

THE MOVEMENT OF THE STARS

No proof in favor of Copernicus's system was more convincing than the discovery of the stellar parallaxes. If the earth is truly in rotation around the sun, then we should be able to see the stars move. Over the course of a year, they should make minuscule ellipses against the background of the dark sky. This shifting, like Mars's retrograde movements, is an effect of perspective: depending on the position of the earth in its orbit, we don't see a given star at the same angle and we have the impression that it's drifting (see sidebar, pp. 152–153). This completely minuscule movement depends on the star's distance: the further away it is, the tinier its parallax will be. In practice, only those of stars closest to the earth can be measured.

FINITE OR INFINITE?

Tycho Brahe's eagle eyes didn't notice any parallaxes. His instruments, the best of his era, were still too imprecise for this sort of measurement, though the Dane had total confidence in them. The absence of parallaxes had two possible interpretations. If Copernicus's model were correct, then this simply meant that the sphere of fixed stars was very, very far from the earth. A gaping void would exist between Saturn's orbit and the fixed stars, a notion that greatly irritated Brahe. The other possible alternative to the absence of parallaxes required purely and simply abandoning the idea that the earth revolved around the sun. This is what Brahe did, for fear of letting the cosmos expand disproportionately. His model, labeled geo-heliocentric, is a compromise between those of Copernicus and Ptolemy: the other planets revolve around the sun, which in turn spins around the earth.

Kepler, or the Fear of the Infinite

Kepler accepted without reservation the Copernican size of the universe. He even pushed its limits beyond the several hundred thousand terrestrial radii advanced by the canon. He still refused, however, to abandon the sphere of fixed stars. The notion that the universe might be infinite anguished him terribly, and the myriads of stars torn from their abyssal depths by Galileo's telescope—who had a weakness for an indefinite universe—didn't succeed in changing his mind. The universe was perhaps enormous, but it remained finite, enclosed in the protective envelope that carried the stars. Kepler was, however, in a good position to grasp all of heliocentrism's implications: if the earth, as Copernicus supposed, is spinning around its own axis, then the stellar trajectories are merely an illusion. There is no longer any need to refer to the movement of a celestial sphere to explain them. Stopped in its tracks, the sky of fixed stars became useless, permitting the stars to scatter where they pleased: finite or infinite, the universe looked the same from an observer's point of view.

The Aberration

After Newton, who strew the stars across the empty and infinite box of space, astronomers gained new interest in the parallax. They wanted to use it to estimate the distances separating various stars (see sidebar, below). The Englishman James Bradley—who would succeed Edmund Halley in the post of Royal Astronomer in

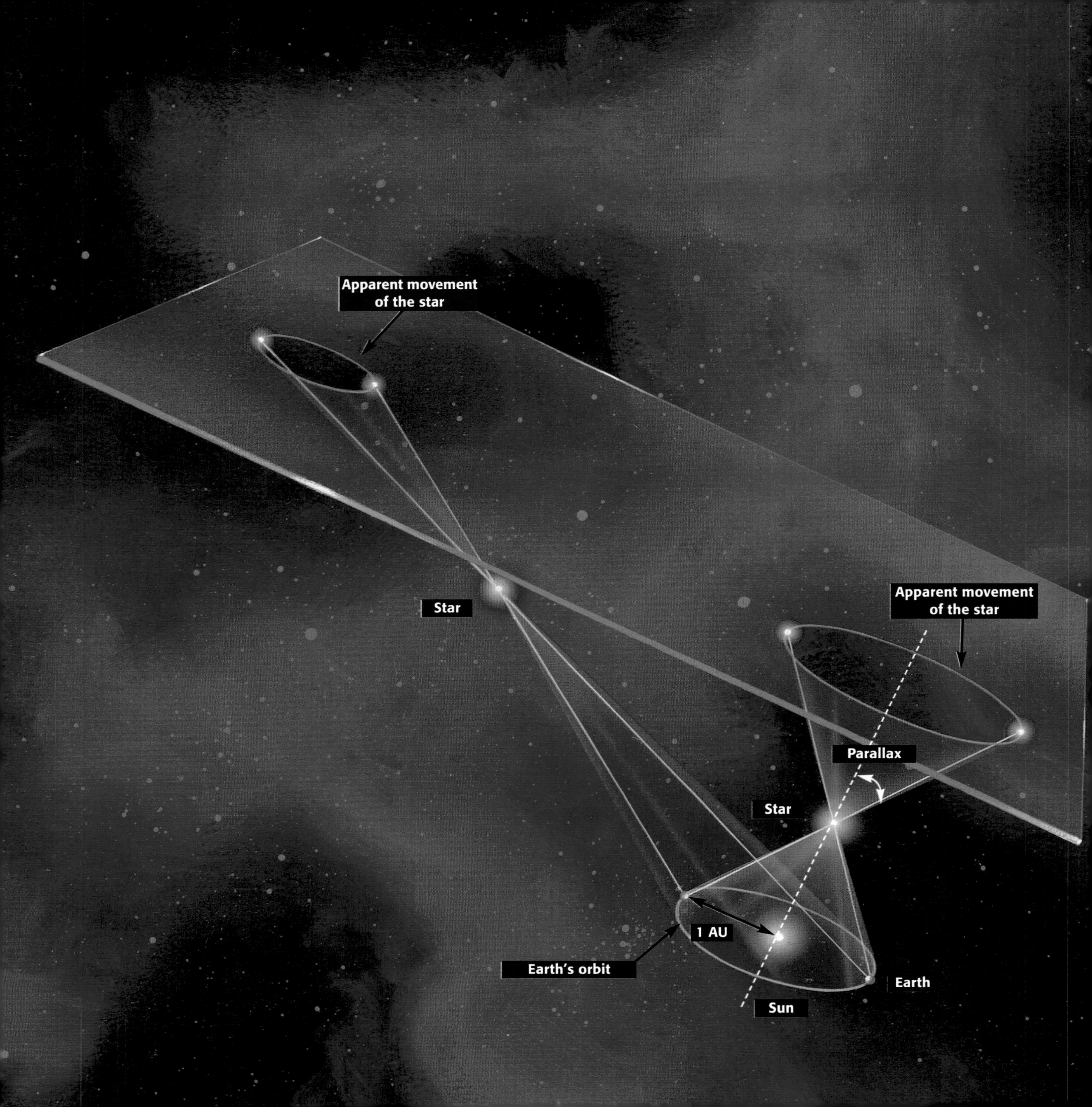

Apparent movement
of the star

Star

Apparent movement
of the star

Parallax

Star

1 AU

Earth's orbit

Earth

Sun

1742—tackled that of the star gamma Draconis in the constellation Draco in 1725. He had a partnership with Samuel Molyneux, a rich amateur living near London, who provided him instruments and facilities. The movement the two succeeded in measuring, a real feat for the era, had absolutely nothing to do with what they were actually seeking. The phenomenon they observed, called aberration, is itself, too, an illusion. The light that reaches us from the stars resembles a shower of

THE STELLAR PARALLAX

Grab a pencil and hold it right in front of your eyes. If you tilt your head to the left and right while keeping your gaze squarely on the object, you'll see it move. This movement, of course, is only an illusion. The pencil's still in the same place; it's the angle at which we're looking at it that has changed.

Replace the pencil with a star and your head with the earth. The latter is moving, and its "point of view" toward the star isn't constant. According to whether it's on one extremity or the other in its orbit, it won't "see" the star in the same place. If, over the course of its annual revolution around the sun, our planet keeps its eyes (our eyes) fixed upon the star, the star will seem to move in a more or less elliptical path.

This apparent movement lets us calculate its distance. If we connect the two lines of perspective joining the two extreme points of the earth's orbit to those of the apparent trajectory of the star, they will cross exactly at the spot where, in reality, the star is located. From a strictly geometric point of view, these two lines will form two triangles, which share a common point—the star. The star's annual parallax will correspond to half the angle at the tip of these two triangles: if astronomers succeed in determining this, then, knowing the earth's distance to the sun, they can deduce the length of the sides of the first triangle. That is, the distance between our planet and the star.

The parallax is the basis for a unit of measurement used uniquely in astronomy, the parsec: this is equal to the distance of a star whose parallax is 1" (arc second). It's worth 206,265 astronomical units, or AU. One AU equals the average earth-sun distance, or around 93.2 million miles. The most familiar unit of measurement is the light-year. This represents the distance traveled by light over the course of one year in a vacuum. Knowing that the latter's speed is 186,282 miles per second, a light-year equals 5,878.5 billion miles.

water droplets fall vertically on our umbrella. If we start to run, we have the impression that the rain is falling at an angle and coming towards us from the direction in which we're hurrying. Because of the earth's movement, light rays seem to emerge from various positions held alternately by the star which, in reality, hasn't moved a hair.

The Size of the Universe

The first parallax, that of star 61 in the constellation Cygna, located eleven light-years from the earth, was discovered by German astronomer and mathematician Friedrich Wilhelm Bessel in 1838. Many others would follow, and new surveying methods based on the analysis of the light emitted by stars would push the limits of the universe to over ten billion light-years from the earth.

FAREWELL TO THE FIXED STARS

In 1718, Edmund Halley recalculated previous positions—measured by Hipparchus and Ptolemy—of the stars Sirius, Procyon and Arcturus. He compared them to data from more recent catalogues, such as one by John Flamsteed, which the horrid Newton had had published without the author's permission. They weren't the same: the stars had apparently shifted in relation to the ecliptic between the second and eighteenth centuries. Halley was sure this movement wasn't due to any change in the plane of the terrestrial orbit, and concluded that the stars themselves must be moving. The change wasn't enormous, only a few arc seconds per year. In 1783, William Herschel discovered that the sun was moving, too. It was moving, with all its baggage

speed of 379 million miles per year toward the constellation Hercules. It's spinning around the center of our galaxy, the Milky Way, which itself is moving. All the galaxies are moving further away from each other, and no one knows when their universal flight will come to an end.

EPILOGUE

Designed to keep up appearances, the sphere of fixed stars will continue to serve as a guide to all those who, some day or other, choose to follow the stars in their age-old wanderings. They will always have the impression of standing on an immobile earth at the center of a moving sky. They will continue to see the stars spin above them, and the planets roam amidst the imaginary characters of the zodiac. The fixed stars will remain fixed, and the Big Dipper, a giant ladle. The heavenly landmarks will remain the same as they were at the time of the Mesopotamians, or of Hipparchus or Ptolemy: the eternally flat dish of the horizon, the hoop and starry belt of the ecliptic, the celestial equator and its axis of rotation. Child of observation, mother of astronomy and irony of history, the sphere of fixed stars has now become an astronomer's trick, a theoretical construction which we must beware of believing, but which nothing prevents us from witnessing.

In 1750, Englishman Thomas Wright published a curious theory of the universe. He imagined an infinite universe populated, not with stars, but with a multitude of cosmic spheres identical to the one in which Plato tried to seal us.
[© Royal Astronomical Society]

PAINSTAKING OBSERVATION OF THE SKY was born with human intelligence and inquisitiveness. The journey of the sun, the phases of the moon, the odd movements of the planets and the passage of comets against the backdrop of the fixed stars have fascinated entire civilizations. The invention of the telescope increased our understanding, pushing the boundaries of the heavens far beyond our mere solar system. This scientific and technological revolution provided us a whole array of amazing instruments; lenses, mirrors and cameras expanded our vision exponentially, unlocking the doors to unknown cosmic realms. For those able to read the program of celestial events, choose their balcony, strap on a pair of good binoculars and find their way amidst the constellations, the spectacle offered by the planets, stars, nebulae and the galaxies will afford endless emotion and delights.

Alain Cirou

KEYS TO THE SKY

ALAIN CIROU

1 Observation with the Naked Eye
2 Tools of Astronomy
3 The Solar System
4 Deep Space
5 Astrophotography

Part Two

1

Observation with the Naked Eye

1. OBSERVATION: AN INSTRUCTION MANUAL 2. FIRST STEPS 3. REVIEW OF THE CONSTELLATIONS

"A sky lit up, a sleeping earth: such are spectacles that free us from this world of noisy passions, pleasures of the soul we can savor in peace. But whatever may be the sweetness of contemplating you, oh beautiful night, however delicious the moments you provide us, the first stars you illuminate across the infinite will always be there to draw us in ever more deeply, to delight our gazes and thoughts ever more dearly." Thus mused Camille Flammarion, astronomer and great chronicler of the night.

Observing the sky with the naked eye is a natural gesture. A sunrise or sunset, a crescent or full moon, a star twinkling at the horizon or a satellite crossing a section of sky are regular events, easy to witness and follow day by day simply by lifting your head. A good lookout point and a small dose of patience will help satisfy your curiosity and familiarize you with celestial phenomena. In astronomy, spectacular events are rare, and their uniqueness only increases when compared to the unchanging rhythms of the humdrum celestial routine they disrupt. Since you must relearn how to see and use all the hidden resources of your eyes, the real pleasure most often comes from direct viewing. Those who learn to contemplate the sky open up a window unknown to the absentminded and hurried. In this discreet and subtle world, the promises are those of infinite discoveries. And they will be kept!

1. OBSERVATION: AN INSTRUCTION MANUAL

As with hikers who equip themselves with good shoes, provisions, backpacks and maps, an expedition into the world of night requires a basic level of preparation.

You must first select an observation site, make sure you're comfortable and minimally secure there, protect yourself from the cold and other climatic irritations, free yourself from all sources of artificial light and, finally, arm yourself with sky maps and a program of the night's events.

CHOOSING YOUR SITE

For a first contact with the sky, it's imperative to find an observation point, a spot far from cities, houses, highways, streets and industrial installations. Slightly elevated and overlooking the surrounding countryside, this "observatory" should be easily accessible day and night and totally clear in all directions.

Light Pollution

Among the nuisances most disliked by astronomers, light pollution is the most common. The most worrisome as well, due to its rate of expansion. If you can easily see a ring of keys lost beneath a streetlamp, there's no point trying to view the stars there: even the faintest light will obscure the stars. Only a full moon can rival their brilliance. In a city or village, public lighting will let you see at best our moon and certain bright planets. You must thus avoid urban centers,

The lights of Earth at night.
[© *Ciel & Espace*/NASA/GSFC/DMPS]

suburbs and highways if you wish to access the stars' faint light in the darkness of night. The brightness of large urban regions requires a distance of more than thirty miles before the luminous dome topping city lights will be reduced to a significant degree. What relationship can we expect to foster with the sky if the stars remain invisible?

Atmospheric Turbidity

Turbidity, also hated by astronomers, results from differences in temperature which stir up the sky and cause eddies. Above an asphalt road heated by the summer sun, the air starts to bubble and disturbs the sharpness of distant objects. In the same way, a concrete terrace, a wall exposed to daytime heat or an open window near the observer will cause the atmosphere's ocean of air to ripple, creating disturbances. When starlight crosses these patches, it dances like the sun's reflection at the bottom of a pool. In contrast, a viewing station set up in the middle of a big garden, a field lined with trees or the edge of a lake will provide calm and stability.

Atmospheric pollution, finally, requires you distance yourself from industrial zones and sources of gas emission and dust: urban zones, highways, and the like. Air stability and purity are, for the observer, precious ecological conditions.

A COMFORTABLE SETUP

Darkness is often cold and damp, and the immobility imposed by keeping the astronomical watch requires fighting against discomfort, fatigue, sleepiness and numbness. The ideal would be to set up a cozy bed in the middle of a meadow.

Sheltered from the Cold . . .

For every problem, there's a solution! A padded, adjustable lawn chair, blankets—at least one safety blanket to totally insulate yourself from humidity and dew—heavy socks and shoes, long underwear, a sweater, padded jacket, thin gloves and a wool cap form the perfect array for night viewing. A minimum of exposed skin for a maximum of heat conserved: that's the underlying formula, summer and winter, for feeling comfortable at night. A thermos filled with tea or coffee, some cookies or dried fruit, moreover, assure energy reserves and complete the equipment.

. . . and from Light

The next step concerns sky maps and the best way to read them without breaking the "chain of darkness." A portable sky map lets you identify constellations, track the major alignments to find and position stars according to the day or hour and, finally, determine their celestial coordinates (right ascension and declination) (see sidebar, pp. 46–47). These maps can be complemented with one of the zodiacal belt, the zone of the sky where the planets, moon and sun travel; they are updated monthly in astronomy magazines. A final preparation before you dive into the night: a flashlight with a bulb covered with red paint or nail polish! This technique was inspired by photo labs, where all light is prohibited. To adapt itself to the darkness, like a cat at night, the human eye needs a good twenty minutes before reaching its maximum acuteness and perceptivity. It's then at its very best performance and can see stars a thousand times fainter than Sirius, the brightest star in the sky. Red light alone will leave your vision unaffected at night: the tiniest light from the flame of

HOW TO READ SKY MAPS

To guide the immobile observer, nailed to the ground in a lawn chair, sky maps do the moving. Like the starry vault around the North Star, these maps revolve around a pivotal point. A movable disk made of cardboard or plastic indicating the hours of the day and the four cardinal points is attached, with a rivet, to a fixed card featuring the constellations and the months of the year. An elliptical window representing the observer's horizon delimits the section of sky visible at a given moment. You need only turn the disk and align the date and hour to locate constellations in the sky. This simplest type of sky map is an image of the celestial vault projected on a 2D surface. It's a good idea to hold it above your head to adjust the depicted sky to the one observed. In practice, and to avoid distortions visible at the map's periphery, you should orient yourself starting from a known constellation. At middle latitudes, locate the Big Dipper in the sky, adjust the date and time on the movable disk, and turn the map so it reflects the sky above you.

Maps of the Southern Hemisphere have no pivot point like the North Star. Rather, the celestial South Pole is occupied by the constellation Octans (named after the octant, a navigational instrument that preceded the sextant). To locate it, you need to imagine it forming the summit of an isosceles triangle whose two other points are occupied by Rigel Centauri, the third brightest star in the sky, and the star Acrux, or Alpha of Crux (the Southern Cross). Beware of time traps! These maps are gradated according to Universal Time (UT): you will need to adjust your star map to reflect how the sky looks in Greenwich, England, at your local standard time. (For example, midnight for an observer on the East Coast of the United States will have to move the wheel ahead 4 hours during Eastern Standard Time, and 5 hours during Eastern Daylight Time.)

Interactive Maps

Digital sky maps that run on PCs and Macs are quality products today and widely available. Installed on a home computer, they let you prepare your observations by combining background maps of stars taken from professional catalogues with ephemerides of the positions of the moon, planets, comets and asteroids. Installed on a laptop, they can be powerful interactive tools at an observation site. Even if the "name," "search," "enlarge" and "zoom" functions seem basic, these visualization commands will largely exceed what you'll see your first time out and help you get beyond weather, space and time constraints. To bone up on astronomical events before a trip, such as getting familiar with constellations in the Southern Hemisphere, the digital tool is ideal. It even lets you observe from Mars or Jupiter an extraterrestrial sky where a certain planet rises and sets . . . Earth.

a cigarette lighter is all that's needed to blind you and force you to start over from scratch.

2.

FIRST STEPS

Astronomy is a matter of patience. This adage, long repeated in popular works, predicts the disappointment often expressed by beginning observers put to the test by their nightly task. Drawn by the excitement of major media events—comets, meteor showers, total lunar eclipses and, more rarely, solar eclipses—and by recent photographs taken by the Hubble Space Telescope, the "little guy" has a good reason to express profound perplexity when given the chance to view the famous spectacle of night directly. Nothing, apparently, seems to move. The stars are stripped of color. The planets look no different from the stars that, no matter how hard they try, make no discernible shapes. The existence of a Lion, Unicorn or Dragon in the sky is as speculative as that of the Loch Ness monster. Nebulae and galaxies, moreover, are easily confused with clouds.

THE FOUR HORIZONS

To appreciate an event at its fair value, you need to know its context so you can grasp all its nuances. On this one condition—establishing a real intimacy with the sky starting from its "little nothings"—a trip into the world of night will never disappoint. To keep from getting lost in the sky, you first need to orient yourself. A horizon, perfectly clear on all sides, lets you determine the four cardinal points.

Conjunction of the moon with Venus.
[© *Ciel & Espace*/P. Parviainen]

West

When night falls, astronomers concentrate all their attention toward the west, which is easy to identify because it's where the sun is setting. Because of the earth's rotation around its axis, the stars all seem to move in the sky. The sun, moon, planets and stars rise in the east and set in the west. To the observer, the illusion's perfect: it's the sun that plunges below the horizon, not the earth rising in the sky! All the stars visible in the west at sunset will quickly disappear, too. The discreet planet Mercury, sparkling Venus—the Shepherd's Star—and the thin crescent of the rising moon are regular visitors to these twilight transformations (see sidebar, pp. 164–165). The planetary conjunctions, those effects of perspective where bright stars, the moon and planets come together in the same line of sight, count among such spectacles.

South

It's in the southern portion of the sky where you can explore the stellar breeding grounds of nebulae and clusters and where the differences in the seasons are most easily distinguished.

The stars all have rendezvous on fixed dates. In summer, it's the flurrying of the Milky Way and the rich constellations of Sagittarius and Scorpio that occupy the southern sky. In winter, these wide spaces belong to hunter Orion and his dogs. These constellations high above the southern horizon clearly mark the periods of the summer and winter solstices.

North

The north is characterized by the permanence of its points of reference. The Little Dipper—whose main star

The sun's journey across the sky.
[© *Ciel & Espace*/P. Parvainen]

is the North Star—is surrounded by constellations noteworthy in that they remain visible above the horizon regardless of the season. These are called circumpolar (spinning around the pole) constellations and, taking Cassiopeia or the Big Dipper as examples, form the basis for getting your bearings in the sky.

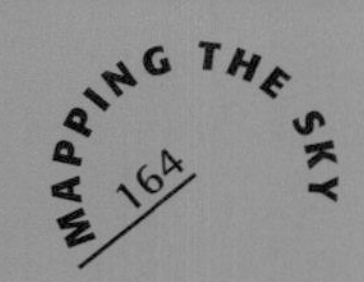

THE CURIOSITIES OF TWILIGHT

And if the atmosphere existed only to trouble astronomers? It's like their very own hall of mirrors, distorting stars, coloring them, creating mirages. Many spectacles that charm naked-eye observers are merely the result of atmospheric games of light. Dawn and twilight are when these luminous fantasies usually take place: the first precedes the royal sun's rise, the other follows its passage below the horizon. They are, however, only two sides to a single phenomenon.

Whether it comes to us from the daystar or distant stellar outposts, light, which spreads out in a straight line through the vacuum of space, is diverted and distorted when it crosses the earth's atmosphere. The curvature and refraction of light rays is all the more pronounced as we descend in altitude. A rather curious consequence of this effect: the sun, moon and stars all appear higher in the sky than they actually are! When the sun sinks below the horizon, it has, in fact, already set a good two minutes earlier. The strange shapes it takes when it sets are the first spectacles of twilight. Again, because of atmospheric refraction, the sun appears either flattened or cut into several pieces, or even shaped like a dented lantern. In very clear and calm weather, on the coast or in the desert, try to observe the mythic "green flash" when the upper edge of the solar disk disappears beneath the horizon. This brief glimmer of intense green is the final color of sunlight refracted by the atmosphere. A few minutes later, the light reaches elevated regions and clouds. Mountaintops are pink in color; purples and violets tint high clouds. Sea spray and dust add infinite nuance to this palette of colors.

At the same time, to the east, a dark and slightly curved mass rises above the horizon. This "antitwilight arch," lined with pink, is the shadow of the earth projecting itself onto the lower strata of the atmosphere. In the mountains, it isn't rare for the edges of this image to bristle with spikes. The evening of the spring equinox and the morning of the autumnal one, on very pure and moonless nights, a privileged few will see an immense cone, shimmering with golden colors. This "zodiacal light," at the threshold of the invisible, is the reflection of the sun's light on the clouds of dust that encircle our planet. Very high up, sometimes more than thirty-five miles in altitude, noctilucent clouds—clouds of volcanic dust—reflect the setting sun's glow even more. Satellite solar panels are the sun's final mirrors and, more than an hour and a half after its fall below the horizon, day gives way to night.

At sunrise or sunset, the atmosphere diverts the sun's light rays and distorts the image of the huge gas ball.

[© *Ciel & Espace*/P. Parviainen]

The glimmer of sunset coloring the upper atmosphere.
[© *Ciel & Espace*/A. Cirou]

Rare and original: the mythic "green flash."
[© *Ciel & Espace*/P. Parviainen]

This is the line below which hide changes to come: the birth of the dawn, the rise of the planets and sun, the appearance of new constellations. Each night, the sky's appearance alters a little bit due to the movement of

Facing north, all the stars seem to spin around the North Star, the practical pivot point located right next to the Pole.
[© *Ciel & Espace*/P. Parviainen]

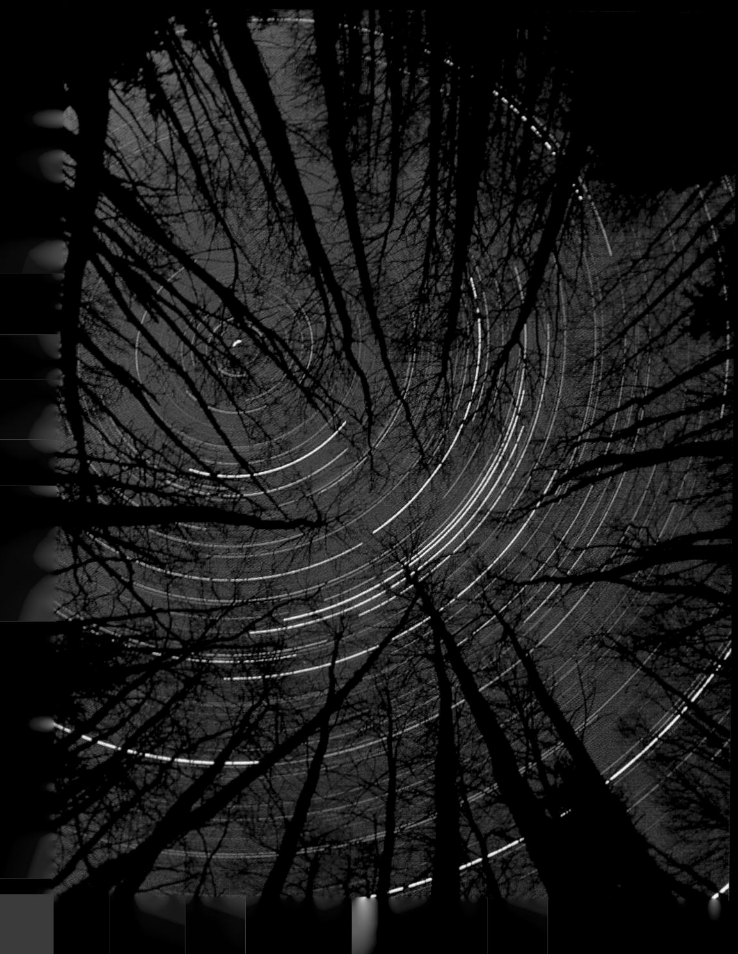

56 minutes and 4 seconds, and completes a full revolution around the sun in one year. The result of this combination of movements is that in the space of a month, the moment a constellation rises on the eastern horizon will shift by two hours! The sky's appearance on 1 January at 2:00 A.M. is the same it is on 1 February at midnight or 1 March at 10:00 P.M. In this way, the rise of new stars on the horizon is a sign marking the changing of the seasons. In the glimmer of dawn, a prelude to the sun's awakening, the appearance of Sirius—the beacon star of the constellation Canis Major—announces the arrival of the August heat (the expression "dog days of summer" comes from their reckoning from the Dog Star's rising). Right behind it with the arrival of giant Betelgeuse, the autumn skies begin chasing away the summer nights. Slowly but surely, the winter stars will follow.

STAR OR PLANET?

Distinguishing a planet from a star is a relatively easy task. At the most, it's like picking five distinct individuals out of a crowd of five thousand.

The Procession of the Planets

In this sparkling and immobile crowd, only the five people in question will be moving. The planets, unlike stars, don't figure on the planisphere because they move. These "errant stars" glide across the starry canvas, with the movements of the fastest among them like Mars and Venus, noticeable from night to night. They always take the same route and frequent the same regions. It's useless to try to get a jump on them by looking north towards the Big Dipper or Cassiopeia

Conjunction of planets Jupiter and Saturn in December 2000, in the constellation Taurus.

the planets all move within the same strip of sky, amidst the zodiacal constellations (Pisces, Aries, Taurus, and so on). It's across this belt, whose center is the ecliptic (the sun's route) that the daystar, moon and planets make their trips across the sky. To the naked eye, Uranus, Neptune and Pluto are invisible, but in this patch of sky, where the principal bodies of the solar system perform their magic and dig their furrows, there's always something going on. Thus continues the music of the spheres.

Mercury and Venus

If all the planets revolve around the sun, the closest ones, Mercury and Venus or the so-called "inferior" planets, are found right on the daystar's outskirts. The planet of thieves and the Shepherd's Star appear in the sky only right before or after sunset, at dawn or at twilight. They are never visible in the middle of the night.

Mars, Jupiter and Saturn

To locate Mars, Jupiter and Saturn, the "superior" planets whose orbits are situated outside that of the earth, keep your eye out for their relative brightness. When they're in opposition, that is, when their distance to the earth is minimal, they can rival the brightest stars in brilliance. Even better, they don't sparkle like the latter and, because of their size—their apparent diameter isn't inconsequential—they aren't especially affected by atmospheric turbidity. Their color is another telltale sign: yellowish white, for Jupiter, slightly orange for Saturn, red for Mars.

3.

REVIEW OF THE CONSTELLATIONS

The keys to the sky are . . . in the sky! To orient yourself without compass or map, to prance from nebulae to galaxies both summer and winter, there's only one solution: get to know the celestial vault, recognize the constellations at a glance, refer to the stars by their nicknames, travel like a tireless foot soldier every path from the zenith to the four horizons—in sum, know your way across the celestial topography like letter carriers learn their rounds.

Becoming familiar with about twenty key constellations is indispensable to feeling at ease during your celestial strolls and, later, to unearthing the secrets of the sky. These constellations serve as a visitor's map, watch and calendar. Their principal stars are universal beacons that serve as points of reference for knowing whether a given twinkling is the explosion of a star or the passage of a meteor. This language of lights is the alphabet of the astronomical dialect.

The Celestial Provinces

The stars making up a given constellation have no common history. Their ages, sizes and distances can be quite different. They're only linked because of their brightness, which causes them to stand out against the background of the sky and lets a rough image be perceived, the symbolic silhouette of a mythological figure. It's useless, however, to look for the figures of a lynx, giraffe or hydra in these imaginary clusters. Principally inspired by ancient Greece, the celestial nomenclature placed in the firmament goddesses, gods and heroes, real and mythical animals and then, more recently, in the sky of the Southern Hemisphere, scientific and navigational instruments (see sidebar, next page).

Left: The great constellation of Orion is the key to the winter sky. The perfect alignment of the three stars that make up Orion's belt, known as the Three Kings, make finding this constellation easy.
Opposite: The constellation of the Southern Cross, key to the southern celestial vault.

[© *Ciel et Espace*/A. Fujii]

BOREAL SKY, AUSTRAL SKY

The boreal sky is the portion of the firmament visible in the Northern Hemisphere. Its primary constellations are well known and seem to spin around a pivotal point occupied by the North Star. For a resident of the Southern Hemisphere, the sky looks quite different. The austral (southern) celestial pole, for example, has no characteristic bright star, and its direction, that of the constellation Octans, is marked by the easily identifiable Southern Cross. Of the two hemispheres, the southern one is, without question, the richer in observable objects. The strip of the Milky Way accessible there is the most dense with nebulae. The two neighboring galaxies of the Magellanic Clouds are universes of billions of stars disguised as hazy clouds. The globular cluster Omega Centauri and its hundred thousand stars, Eta Carina, the "Coalsacks" and the "Jewel Box" are celestial curiosities astronomers below the tropics owe it to themselves to observe. Here, too, a sky map is obligatory for orienting yourself among these strange constellations. From Argo Navis, divided into four constellations—the Compass, the Keel, the Sails and the Poop—to the Painter's Easel or the Sculptor's Studio, a trip across the southern sky remains an unforgettable spatio-temporal voyage.

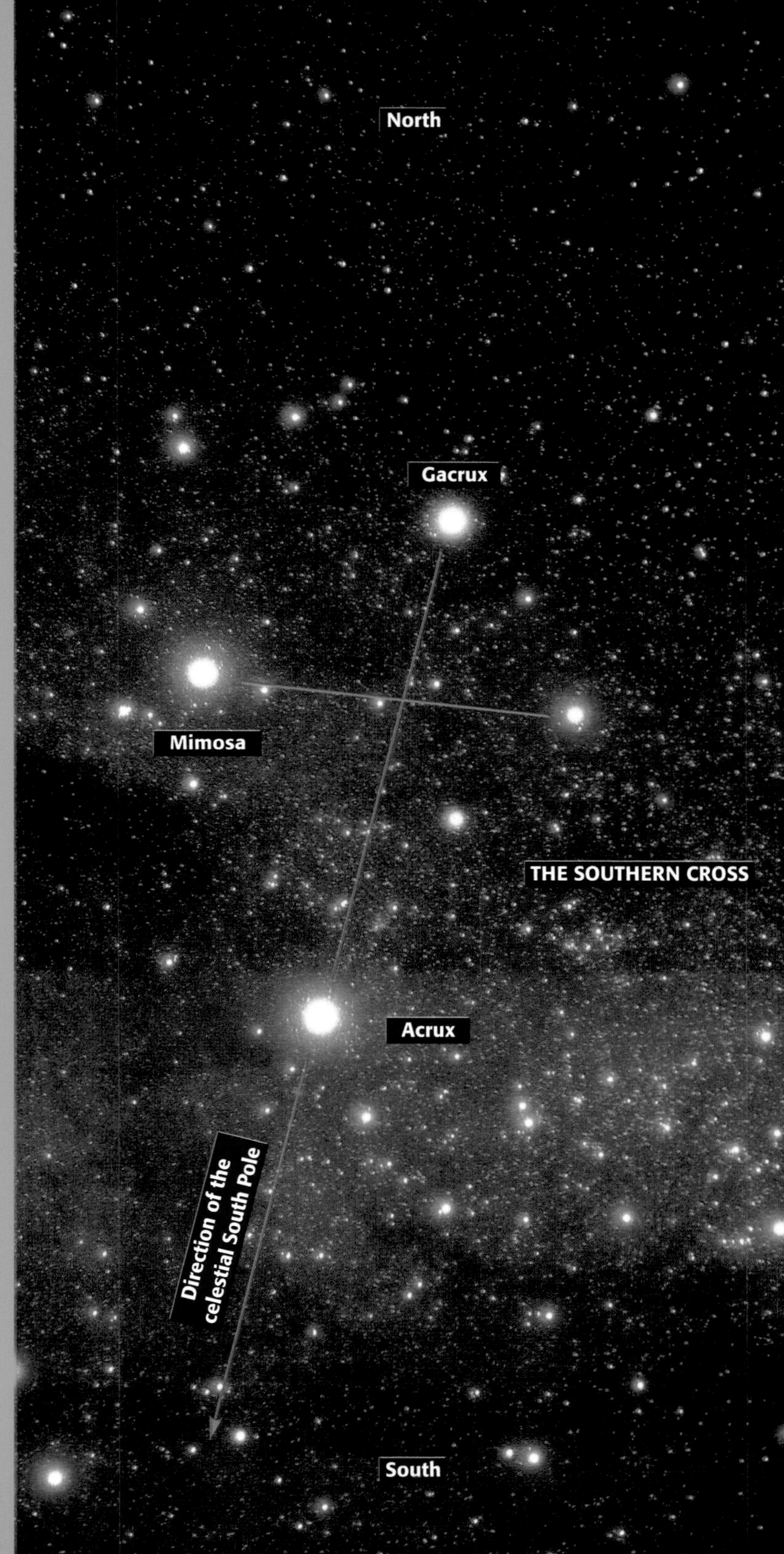

THE MAGNITUDE AND COLOR OF THE STARS

You need only watch the stars light up in the sky one by one to realize that they aren't all equally bright. Based on their differences in brightness, Greek astronomer Hipparchus in the second century B.C. defined six classes, six "sizes" of stars. Astronomers later quantified this simple visual appreciation by creating "magnitudes." According to this logarithmic brightness scale, a star of magnitude 1 is two-and-a-half times brighter than a star of magnitude 2. The faintest stars visible to the naked eye (magnitude 6) are a hundred times less luminous than those of first magnitude. Curiously, the brightest stars are awarded negative magnitudes. That's the case of Sirius (m=–1.44), Venus (m=–4 maximum), the sun (m=–27)—and the light of the full moon (m=–12). The faint galaxies photographed over tens of hours of exposure by the Hubble Space Telescope reach a magnitude of 29: the stars at the edges of the universe are 400 billion times weaker than a star of magnitude 1.

Magnitude is characterized as "apparent" because it doesn't take into account a star's true luster. A nearby star will always seem brighter than its distant twin. To really account for a star's luminosity—in the physical sense of the term, it's the total quantity of energy it produces per unit of time—astronomers refer to "absolute" magnitude: this is the apparent magnitude these stars would have if they were all located the same distance from the earth. Which is, by convention, 32.6 light-years from us (that is, ten parsecs, ten times the distance at which the mean radius of the terrestrial orbit subtends an angle of one arc second). The sun, which obviously looks far brighter to us than Sirius, has, in reality, a twenty-five-times fainter luminosity. Sky atlases—more specialized than sky maps—publish these two values as well as the stars' distances in light-years.

To our eyes, the sky looks nothing like Christmas tinsel, and its little lanterns are hardly distinguished by color. However, looking at them more closely in clear and calm weather, the brightest stars emit red, orange, yellow or blue glints. This subtle sparkling that hits the cones of our retinas—the cells sensitive to color—reveal the temperatures reigning on the stars' surfaces and provide a fair indication of their respective ages. The blue Sirius, in the constellation Canis Major, is very hot (more than 36,000°F) and young. Auriga, the Charioteer, is, like the sun, yellow and in its prime, with a temperature of 10,800°F. Arcturus of Boötes is an old, orange star and the red Antares of Scorpio, ten thousand times more luminous than the sun, is a super giant of respectable age whose color is often compared to that of planet Mars.

At nightfall, the brightest stars, those of first magnitude (see sidebar, p. 170), help delimit the principal contours of the celestial provinces. It's the most favorable moment for appreciating their size and surface area and comparing, with the help of sky maps,

Starting from the seven principal stars of the Big Dipper, we can draw imaginary lines linking all the constellations together.
[© *Ciel & Espace*/A. Fujii]

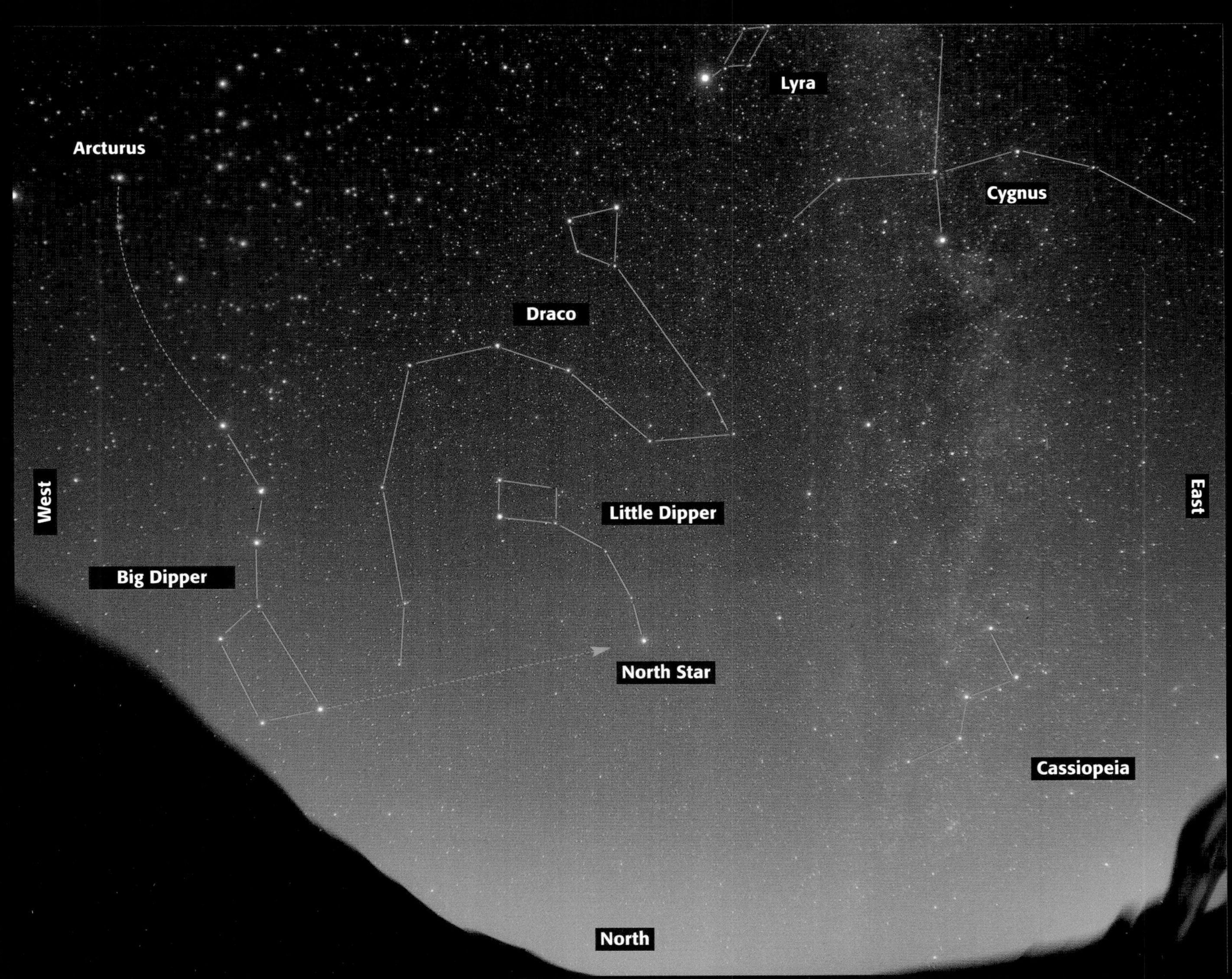

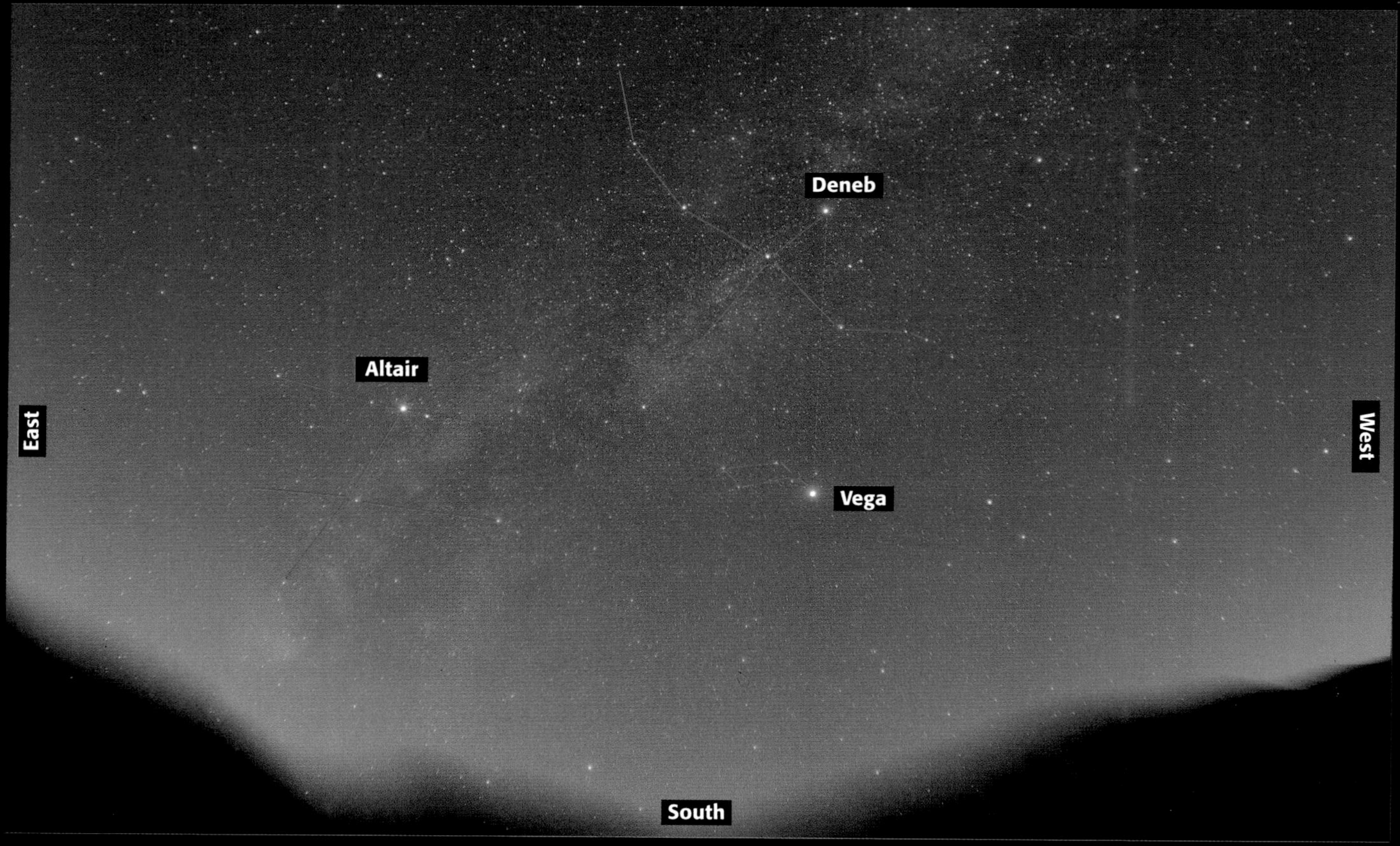

Deneb
Altair
East
West
Vega
South

Deneb of Cygnus, Vega of Lyra and Altair of Aquila are the three stars that form the Summer Triangle.
[© *Ciel & Espace*/A. Fujii]

drawings to reality. It's always surprising to discover that they occupy a considerably wider surface of the celestial vault than the map will lead you to believe. You just have to accept that a simple line drawn between two stars sets the limits of a constellation like the Hunting Dogs, or that three luminous points make up Berenice's Hair alone.

The Big Dipper

If you remember that the sun sets due west at the beginning of spring and autumn, you need only make a quarter turn to the right of the spot where it's just set to find north. Looking upward, it's next quite easy to locate the seven stars of the Big Dipper in the shape of a ladle, or a chariot pulled along by a team of three horses. Keystone of the constellations, it possesses the double quality of remaining visible permanently at our middle latitudes and serving—with Orion for the winter sky—as the center of the major alignments on summer celestial strolls.

The Milestones of the Night

By linking the stars together using imaginary lines, it's easy to find your way amidst key constellations. For example, taking the Big Dipper as your starting point, a line drawn between the two stars beside the ladle lets you find the North Star and the Little Dipper. Shot much further, this virtual arrow points out the gigantic constellation of Pegasus and its accompanying procession: Andromeda, Lacerta and Aquarius. These lines aren't always straight: continuing in the direction indicated by the three stars forming the ladle's handle, a curve will lead to the star Arcturus (the Bear's Arc) in the constellation Boötes, then to the Ear of Virgo (Spica). The famous "summer triangle" composed of the stars Vega of Lyra, Deneb of Cygnus (the furthest star visible to the naked eye: three thousand light-years away) and Altair of Aquila is an excellent passport for crossing, one by one, the borders of the constellations surrounded by the Milky Way. In winter, a line drawn between the Three Kings of the constellation Orion allows you to identify Sirius, the brightest star in the sky. Continued in the opposite direction, it joins Aldebaran, the beautiful red giant in Taurus, then the clusters of the Hyades and Pleiades. Examples like these are legion. Like our ancestors were several millennia ago, we're still free to trace our own routes, to find our own shortcuts using the sky's most obvious landmarks.

OBSERVING ECLIPSES

Solar and lunar eclipses are two spectacular manifestations, casting entire worlds into shadow before pulling them back into the light.

The Solar Eclipse

Through an extraordinary coincidence, the moon and sun, seen from the earth, have the exact same size. In reality, the sun is four hundred times larger than the moon, but also four hundred times further away. When our satellite passes right in front of the sun, it blocks it out completely: this is what's called a total solar eclipse. The phenomenon isn't rare, but to see it you may have to travel far due to the narrowness of the strip of total darkness. If possible, try to position yourself at the center of the lunar shadow's cone to get the full benefit of the spectacle.

Equipped with protective glasses recommended by experts in solar safety, the witness to a total eclipse will be surprised by how slow the process is for the black mask of the moon to superimpose itself totally over the solar disk. Slow as molasses, a slight indentation nibbles away little by little the sun's luminous face. You'll have to wait a good hour before you'll notice subtle changes in the surrounding landscape. The patches of light filtered by the leaves of trees begin projecting clear crescents on the ground. Shadows and colors start to change. Blue, gray and wintry silver tones now predominate. Such pallid flashes give the shadows greater contrast and accentuate the feeling of coldness.

A few minutes before the eclipse is total, "flying shadows" will start spreading through the air, across the ground and over light walls. They're the reflection of moving air crossed by the final rays of the solar crescent. Birds and farmyard animals will suddenly withdraw. In this dramatic landscape, lit 360° by the glimmer of sunset, the spectacle then begins.

Approaching from the west, the gigantic shadow's brush extinguishes the remains of the day like blowing out a candle. In the sky, a rosary of brilliant pearls—Baily's Beads—bring out the irregularities in the lunar circumference. The sun's final burning embers are smothered in a strange fire. Sparks of ruby red gas spurt from the edges of the now blackened star. These protuberances, immense arches of white-hot hydrogen, reveal the presence of eruptions on the solar surface. The solar corona, a silvery glory which circles the star like an atmosphere, affords it a halo as brilliant as a full moon. A multitude of luminous spurts, the "Coronal Jets," shoot out toward the poles like the ears of a sheaf of wheat. In the surrounding space, Venus and Mercury can be seen shining, as several stars stand out against the night-blue background. After several minutes of eternity, filled with the irrational fear of never seeing light again, a glint of molten metal suddenly pierces the darkness in a quick flash, bringing the total eclipse to an end. A moment of incredible grandeur.

The Lunar Eclipse

The conditions for observing a total lunar eclipse are quite different. The three heavenly bodies are again perfectly aligned, but it's the lunar disk's turn to move into the cone of the earth's shadow and be swallowed up there. A popular phenomenon, the eclipse affects an entire half-earth and counts several billion people among its spectators. Slowness is still the rule, as a good hour is necessary before the shiny disk of the full moon disappears beneath the

A total eclipse of the moon occurs when our natural satellite crosses the cone of the earth's shadow. The gleams of sunset then illuminate its surface.

[© *Ciel & Espace*/A. Fujii]

shadow of the earth. This occurrence provides the unique chance to discover, without ever leaving our planet, that it's actually round! At the moment when the final crescent of the moon is erased, the earthly landscapes darken and all light disappears. A modest, colored shine, ten thousand times weaker than normal solar illumination, reveals our satellite's discreet presence. From the moon's surface, observers would see their own solar eclipse unfold, due to the earth. As the latter is surrounded by an atmosphere which filters and deflects solar light, colors comparable to those of our sunsets illuminate the lunar reliefs. Bronzed, dark, deep red, orange or blood red, the colors of the disk of Selene will depend on the transparency of our atmosphere and the degree of solar activity. Perfectly safe, observation of the phenomenon can be done with a good pair of binoculars. Their luminosity will capture all the subtleties and nuances of the "red eclipse," games of light and atmosphere whose secret is kept by the stars.

There's a total eclipse of the sun when the moon masks the disk of our star. From several seconds to several minutes, the halo of its corona is visible.

[© *Ciel & Espace*/F. Espenak]

2
Tools of Astronomy

1. BINOCULARS: POCKET EYES 2. TELESCOPES: NIGHT VISION 3. TELESCOPES: LIGHT TRAPS

"What's essential is invisible," the Little Prince confided to poet-pilot Antoine de Saint-Exupéry. The temptation is great (perhaps because of a little planet inhabited by a rose) to extend this remark to the whole celestial vault. If, on the purest of nights, the sky seems speckled with an infinity of stars, this spectacle is only an illusion! At best, on a dark, clear and moonless night, 2,600 to 3,000 stars can be seen by the naked eye. In the parks and gardens of a big city, this estimate drops to around 200. Knowing that our galaxy, the Milky Way, contains between 100 to 300 billion stars and that there exist billions of similar ones throughout the universe, the window directly accessible to us pales in comparison.

Lenses and telescopes have singularly expanded our universe. Worlds appear nearer and more numerous, and astronomers—using intervening tools—have access to unsuspected riches. With a pair of binoculars, the population of visible stars surpasses the hundred thousand mark. An amateur telescope of a 60 to 250 mm diameter quickly provides access to millions of lights throughout the night. Artificial eyes, giant lenses whose opticians today produce cyclopean examples thirty feet in diameter, cast their gaze without blinking billions of light-years away.

These instruments are veritable funnels: they trap the most light rays possible emitted by the stars and concentrate them into a single point. This is where an image of the stars is formed, which observers can admire to their hearts' content only by keeping their eye against an eyepiece.

Astronomers' tools have long been expensive and reserved for an elite. Today, their access has been democratized, and the technical progress—optical, mechanical and digital—that let this revolution occur led to the development of recreational astronomy. To find the shoe that fits, you should take into account certain criteria to determine your ideal instrument: age, experience, mobility, budget, bulk, nature of the observation site, type of objects viewed, and so forth. All work together to define your dream telescope's silhouette—a serious and important choice, which will set the tone for your voyage amidst the stars.

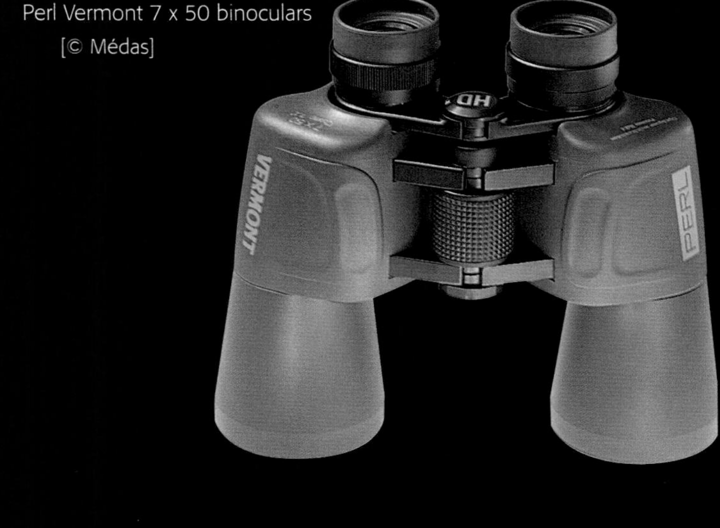

Perl Vermont 7 x 50 binoculars [© Médas]

1.

BINOCULARS: POCKET EYES

Binoculars are the bare minimum needed to penetrate the world of night and hunt its biggest stars down like paparazzi. They have many advantages. First, they let you observe all of nature's spectacles, whether they take place during the day or at night. Spying on the comings-and-goings of a nest of swallows or watching the setting of a lunar crescent both represent classic services provided by this instrument: satisfying your curiosity at a distance. Next, they're easy to handle and not at all bulky, and can easily slip into a glove compartment or bag. They're also very mobile, readily available and discreet. Finally, because there exists a wide range of binoculars including quality models—very luminous ones—especially designed for astronomical observation, these pocket eyes never disappoint by what they show and always inspire by what they suggest.

LUMINOSITY AND DEPTH

The major asset of a good pair of binoculars is luminosity, that is, the number of faint stars it lets you see. Luminosity is indicated by the two numbers etched on the frame: 7 x 50, for example. The first figure specifies the magnification—seven times—and the second indicates the diameter of the lenses (the binoculars' two big eyes) in millimeters. Dividing the latter by the magnification (50 millimeters/7 = 7.1), we obtain the value of the exit aperture. The closer this is to 7 millimeters—the theoretical value of the dilation of the human eye in total darkness—the more luminous the binoculars. Our eye is incapable of competing with a simple 60 mm diameter telescope, which is a hundred times more luminous.

The most current models include 8 x 30, 10 x 50 or even 7 x 50 and 11 x 80. The first aren't especially luminous (like theater glasses, whose field and magni-

astronomical observation, provided you install them on a tripod to facilitate handling. The larger the lenses, the more light they collect. Their weight and price follow the same progression, and certain astronomical binoculars of large diameter—such as those of 20 x 100 or 40 x 150—compete with telescopes. They offer a binocular vision quality very appreciated by dedicated observers. With such binoculars, the image of a large comet deploying its gaseous plume provide depth and relief, a "three-dimensional" effect which no other glass pair of eyes are capable of offering.

Vixen 30 x 125 binoculars.
[© Médas]

WHY BINOCULARS?

Binoculars will familiarize you with the techniques of observation, finding points of reference and exploring entire stretches of the celestial vault. You mustn't forget that the instrument is designed to serve its user. To know it, handle it in the dark and adjust it for viewing, is to tame and take possession of it. To reduce fatigue and avoid shaking, binoculars can be mounted on a photographic tripod thanks to an adapter. Once the stars are in focus—they should be fine and point-shaped—the voyage can begin.

Wide-Open Spaces

There doesn't exist a rule for navigating the skies by sight. Some people like to lose themselves to be better surprised by what they find, while others, in contrast, prefer to plan their itineraries on sky maps and proceed toward specifically chosen targets. The major advantage of binoculars relates to the vast celestial panorama to which they provide access. The observation field, the portion of the nocturnal vault that can be seen using any instrument, binocular or telescope, is inversely proportional to its magnification coefficient (see sidebar, p. 182). The more the instrument magnifies, the smaller its field will be. For the astronomer, a telescope corresponds to a telephoto lens and a pair of wide-angle binoculars. With 7° of field, or a surface equivalent to fourteen full moons for 7 x 50, and a bit more than 4° for 11 x 80, they open the equivalent of a picture window onto the night. The most beautiful spectacles to see are unquestionably galactic clusters and nebulae—Orion for the winter sky and the America nebula for the summer one—subtle clouds that stand out against the perfectly black background of night.

High Technology

The array of astronomical binoculars has benefited from recent technological advances. New high refractive-index lenses improve optical quality; O-ring joints and an inert gas—nitrogen—injected directly into the body of the binoculars increase airtightness and protect against dust and moisture. Certain models are equipped with optical sensors to guarantee perfect focusing. Others, and this is not the least of their perfections, are equipped with image stabilizers to compensate for shaking due to fatigue. Gyroscopes detect chaotic jostling and vibrations; a mobile optical unit corrects such movement. The effect is spectacular and the feeling of comfort is maximal. Finally, on the high end, a wide range of accessories including eyepieces, doublers, sensors and tripod adaptors, raise binoculars to preferred status when you're looking for a companion for those first nights out.

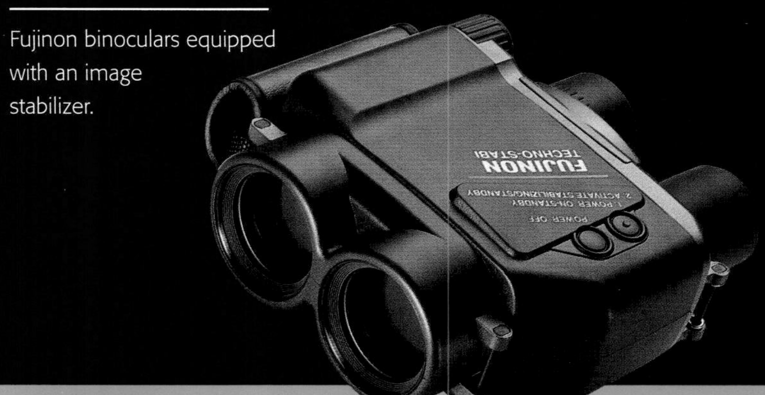

Fujinon binoculars equipped with an image stabilizer.

HOW MUCH DO THEY COST?

For an initial estimate covering the purchase of a pair of type 7 x 50, 8 x 40 and 10 x 50 binoculars, the minimum budget will be around $80. Designed for both astronomical and terrestrial observation, these instruments will give you an excellent idea of the pleasures of celestial exploration. From $80 to $160 the choice is wider, the binoculars less bulky and lighter with superior mechanical and optical performance.

From $230 to $800, numerous models combining maneuverability with remarkable optical qualities are available (including the famous Perl 12 x 80). Subtle—but important—details such as image contrast, edge-of-field definition (these are the outermost edges of the portion of sky visible through binoculars) are basic determining factors. The stars don't lie: they remain perfectly point-shaped, whatever magnifications you use. If, at the edge of the field, starlight becomes distorted like a hang glider or comet, only one verdict can be reached: the optical apparatus isn't perfect.

From $900 to $1,500, manufacturers offer very luminous binoculars (from 8 x 50 to 14 x 100), with impeccable, guaranteed, zero-defect optics. In this price range, the consumer favors comfort, durability and design, and is banking on an instrument that will remain in excellent condition for a good number of years.

Beyond that, for 20 x 100 or 30 x 125 binoculars with exceptional performance, a budget reaching $4,600 may be necessary. Presented on metal columns because of their weight, these instruments are marvels for the wealthy amateur. There exist 25 x 150 and 40 x 150 models, but they can cost over $15,000, an astronomical budget for undeniably fascinating binoculars.

2.

TELESCOPES: NIGHT VISION

High on its haunches, its barrel pointed toward the sky, the astronomical telescope has a familiar silhouette. Developed in 1608 by Dutch eyeglass-maker Hans Lippershey, aimed heavenward for the first time by Galileo, its principles were laid out several years later by Johannes Kepler. An astronomical telescope is a refractor that functions according to a few simple optical laws. It's a rigid and empty tube, on the front of which is placed a lens (an objective) made up of several lenses that deflect (refract) light rays emitted by stars and cause them to converge toward the rear at the so-called focal point or focus. This is where the image of the observed object is formed. At the other extremity of the tube, a magnifying glass (an eyepiece) is placed upon a movable device (the focuser) that serves to enlarge the image (see sidebar, p. 182). A viewfinder—a small lens that enlarges from five to ten times—is attached to the tube. It's equipped with a cross hairs (a reticule) that permits exact pinpointing of a star. A defogger is placed in front of the lens. All of this is installed on a rigid mount designed to facilitate horizontal and vertical shifting, guarantee stability and favor slow tracking (see sidebar, p. 192).

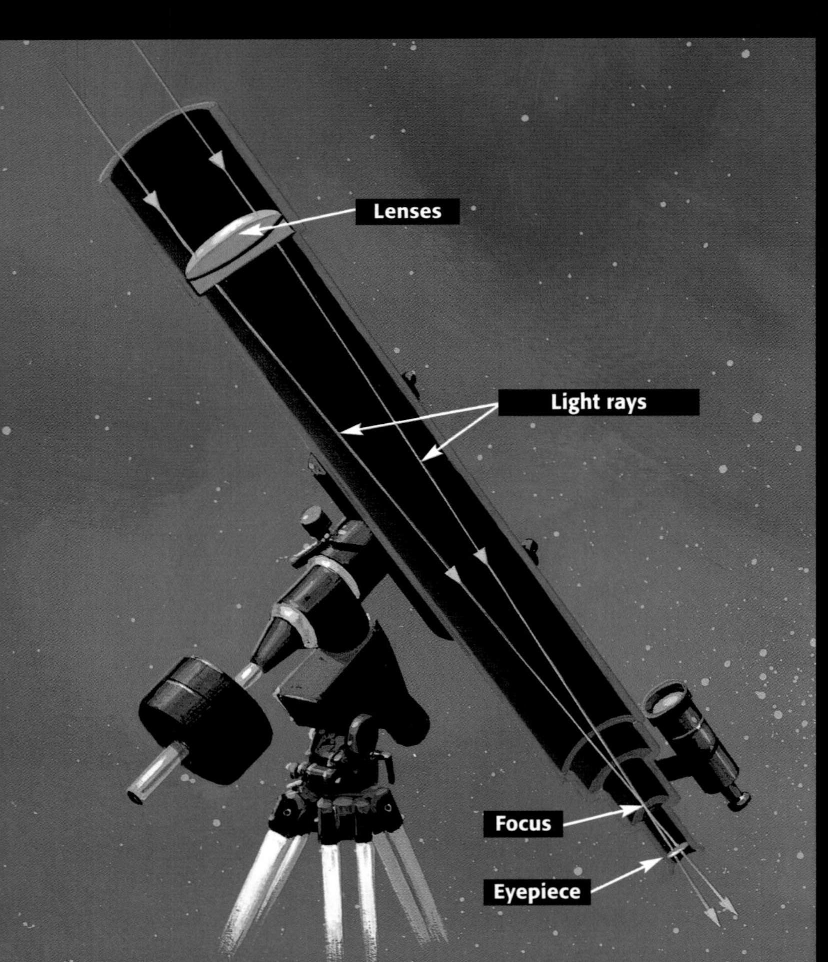

LIGHT AND OPTICS

The lens glass used in the first telescopes scattered light like a prism. This defect, called chromatic aberration, adorned the stars with rainbow colors and distorted images heavily except at the center of the field. This is how Galileo managed to draw ears on Saturn. These problems, which drove Isaac Newton to devise a telescope in 1668, found their solutions. The first involved the creation of slightly domed (convex) lenses, which strongly increases the distance between the focal point and lens. The inconvenience: bulk. The largest telescope of this kind reached 125 feet long! The second

Principle behind the astronomical telescope.
[© *Ciel & Espace*/Manchu]

solution combined two lenses with surfaces of different curvatures. This is the system used today in all small commercial telescopes that are labeled achromatic. Progress in optical formulas—both in the nature of the glass used and the number of lenses combined—has led to the creation of a second type of instrument with nearly perfect properties. These telescopes, described as apochromatic, offer clear, fine and high-contrast images. They're easily distinguished from the rest: they're the most expensive.

CHARACTERISTICS OF TELESCOPES

Three elements characterize an astronomical instrument and are indisputable criteria for choosing: the diameter of the lens, the focal length and the aperture ratio.

Diameter

The diameter (D) of a lens, which is measured in millimeters, provides information about the quantity of light collected and the sharpness of the smallest perceptible details. The greater this value, the greater the number of faint stars will be accessible to the observer. For an initial purchase, a 60-millimeter lens—which lets you admire Saturn's rings and details of the lunar craters—is recommended.

Focal Length

The focal length (f) is the distance separating the lens from the focus, that is, the place where all the light rays converge to create a sharp image. The greater this value, the larger the image will be. On the other hand, the telescope will be less luminous.

RESOLVING POWER

The resolving power—or resolution—of an instrument is its capacity to distinguish between two points very close to one another. It corresponds to the minimum distance separating two bodies below which they'll only appear as one. Most double stars cannot be seen because the angular distance separating them is too small: to the naked eye, their light blends together into a single spot. At its best, the human eye's resolving power is equal to one angular minute, or the diameter of a quarter seen from a distance of three hundred feet (0.5° or 30 arc minutes represent the diameter of the full moon in the sky; one arc second, a detail of 1.14 miles in surface area).

To calculate, in arc seconds, the theoretical ability of an astronomical instrument to distinguish between two points, you need only divide the number 12 by the diameter of the lens expressed in centimeters. Thus, a pair of 50 mm binoculars supposedly can tell the difference between two distant stars of 12/5 = 2.4 arc seconds; a telescope of 20 cm separates an intimate stellar couple of 12/20 = 0.6 arc seconds. The greater the diameter of the telescope's lens or mirror, the smaller its resolving power will be—and therefore, the better the instrument. So much for theory.

In reality, atmospheric turbidity plays the spoilsport. When the atmosphere is stirred up, masses of air affect the sharpness of stellar images. Under the best skies, a telescope, whatever its diameter, will never get beyond a resolution of 0.3 arc seconds. Even then, that's exceptional!

MAGNIFICATION

In theory, the greatest possible magnification an instrument can offer is determined by multiplying the diameter of its lens or mirror by a coefficient of 2.4. A 100 mm diameter telescope will magnify a maximum of 240 times. In reality, accounting for observation conditions, it's more reasonable to expect a value between 1 and 1.5. A small 60/700 telescope will enlarge between 60 and 90 times. And that's still quite a bit! Careful: the greater the magnification, the more limited the field of sky the instrument will cover.

Seeing Double or Triple with Barlow

The Barlow lens is a clever optical device slipped inside the focuser over the eyepiece. Its function is to double, or even triple, the focal length of the lens and thus increase proportionally its magnifying capacity. A high-quality Barlow lens offers a real advantage to observing details of bright bodies, like the moon or planets, under good conditions. The loss of luminosity (four times less light per doubling of magnification) and of field considerably reduces the value of these lenses when trying to observe nebulae and galaxies.

Aperture Ratio

Finally, the aperture ratio (f/D) is the focal length divided by the diameter, expressed in millimeters. For example, a telescope 60 mm in diameter with a 900 mm focal length is said to have an aperture of 15 (or 900/60). This data will give you an idea of the type of targets favored by the instrument. Thus, an f/D superior to 10 is perfectly suited for lunar and planetary observations requiring high resolutions (see sidebar, p. 181). With lesser values, the instrument's performance is ideal for stellar and galactic observation. It then offers a wide field and strong luminosity for objects where maximum magnification isn't required.

Left: Perl equatorial telescope, 60 mm.

Opposite: Perl-Vixen equatorial telescope, ED 106/660.

[© Médas]

Magnification

As surprising as it may seem, magnification isn't a telescope's most important quality. The advertisement praising "the telescope that magnifies 200 times" is using the same specious line of argument according to which a computer calculates "a hundred thousand times faster than Einstein" or light moves 27 million times faster than Ben Johnson. At the focus of the largest telescope in the world, the image of a star remains no larger than a pinhead. If it's far more luminous than in a small, 60 mm telescope due to the amount of light rays collected, its distance to the earth still remains too great for it to have even the slightest

HOW MUCH DOES IT COST?

Starting off with a small telescope will require an investment of at least $400. For less than that, models sold by mail order or in high volume are more toys than real instruments: the optical apparatuses are made of plastic, the stands unstable, the drive mechanism derisory and illusory. Worse: such junk risks turning you off to observation permanently! It's better to stick to specialty stores—the salespeople are almost always amateur astronomers—that offer a wide range of quality models.

From $400 to $600, telescopes of 50 to 60 mm in diameter will reveal the lunar surface and the planets. These instruments are limited, hardly upgradeable, but provide a good introduction to pointing and observing. From $800 to $1,200, small, compact apochromatic telescopes of 70 mm diameter for 480 mm of focal length are available. Though of very high optical quality, they're sold without tripod or mount. Their low bulk is ideal for traveling. From $1,200 to $2,000, a fine collection of telescopes ranging from 80 to 100 mm diameter, installed on stable and mobile equatorial mounts (see sidebar, p. 192) are the cream of the crop. A great number of optical and photographic accessories will complement your initial equipment.

The more you love it, the more you'll want to spend! The price of a high-end apochromatic astronomical telescope of 100 mm verges on $4,000. Beyond that, for diameters of 120, 150 or 178 millimeters, expect to spend anywhere between $7,000 and $40,000.

disk-shaped appearance. An instrument's maximum magnification is not the first criteria for choosing it, but rather its diameter, luminosity and quality of its optics and accessories. What matters most is the eyepiece, for its primary function is to enlarge the image formed at the focal point (see sidebar, p. 189).

WHICH ONE TO CHOOSE?

Astronomical telescopes are like the rungs on a ladder. They let you move from the closest objects to the most distant ones, and are numerous enough and of sufficient quality to let you explore all the side trails of the celestial vault. Beginners who hope to get comfortable with handling and pointing an instrument, to discover—under far better conditions than Galileo did—the upper reaches of the solar system and to practice stellar observation, will top their first mount with a small telescope. An astronomy-lover and city-dweller with access to a balcony or small garden can easily pick out Mars, Jupiter and Saturn and observe lunar formations with a telescope of 80 to 100 mm. Compact, easily installed and stored in their cases, these medium-diameter instruments are sturdy and only rarely break down. Due to their sealed tubes, they're hardly sensitive to instrumental turbulence. Finally, at the top of the scale, the proud owner of an apochromatic telescope (ED, Super ED or fluorite lenses from 100 to 180 mm diameter) will benefit from an exceptional instrument combining strong magnification with impeccable imaging, tripod stability, mechanical photofinishing and design that only the top manufacturers of scientific instruments know how to achieve. A telescope for the price of a boat or nice car . . . now that's worth a look!

3.

TELESCOPES: LIGHT TRAPS

The telescope is to the infinitely large what the microscope is to the infinitely small: a keyhole peering out upon a universe of inaccessible scale. Required accessory for caricatures of eccentric scientists, the reflecting telescope is above all an original, adaptable and decisively modern scientific instrument. Invented by young Isaac Newton, it was presented for the first time in 1671 to British monarch Charles II. This first reflecting telescope had a metallic mirror made from a high-grade copper alloy measuring 3.8 cm in diameter. Its inventor confirmed having "clearly seen Jupiter's globe and its satellites." More than a century later—around 1850—French physicist Jean Foucault experimented with the first glass mirrors and, in 1908, thanks to the talent and tenacity of the American George Hale, an example of 1.5 meters gazed out upon the night from the Mount Wilson observatory. Today, the telescope has transported us billions of years in space and time to the most faraway galaxies in the cosmos.

ANATOMY OF A REFLECTOR

The reflecting telescope is a tube—much more stocky and wide than the refracting telescope—at the end of which is installed a round mirror. This mirror is concave and polished extremely carefully until it takes the shape of a parabola. A fine layer of reflective aluminum is placed upon on its surface. The light coming from the stars enters the cylinder and strikes this mirror, called the "primary." Its particular shape reflects light

rays and causes them to converge without dispersing at a focal point located at the front of the telescope. This luminous bundle is then intercepted by a flat, secondary mirror set inside the tube. The secondary mirror diverts the light a second time, causing it to leave the telescope and strike the eye of the observer.

Reflecting telescopes possess the same qualities as refracting telescopes: equivalent luminosity, aperture ratio, magnification and resolving power. These four parameters depend on the diameter of the mirror and the focal length of the telescope (that is, the distance separating the primary mirror from the point at which light rays converge).

Reflecting Telescope versus Refracting Telescope

Reflecting telescopes use mirrors, while refracting ones depend on lenses. The first formula has clear advantages: mirrors don't disperse colors like lenses do and offer a totally achromatic image, that is, without any rainbow effect. They're cut as one whole piece and don't require glue in their manufacture, additional assembly, or the careful adjustments of lenses. They are available in a wide variety of diameters. Beyond one meter in diameter, the lenses of large, professional refracting telescopes distort light under their own weight. The less cumbersome mirrors of reflecting telescopes are easier to maintain and can be lightened even more by a clever honeycombing process. The larger they are, the more light they collect and the fainter the stars they let you see. Similarly, the resolving power of reflecting telescopes, their capacity to distinguish between objects very close together, increases with the size of the mirror (see sidebar, p. 181).

TWO GREAT FAMILIES

There are many kinds of telescopes: Newtons, Schmidts, Maksutovs, Ritchey-Chrétiens, Cassegrains, Schmidt-Cassegrains, and so on. To each of these names correspond different optical systems, mechanical components and purposes. The Newton and Schmidt-Cassegrain types are the most popular.

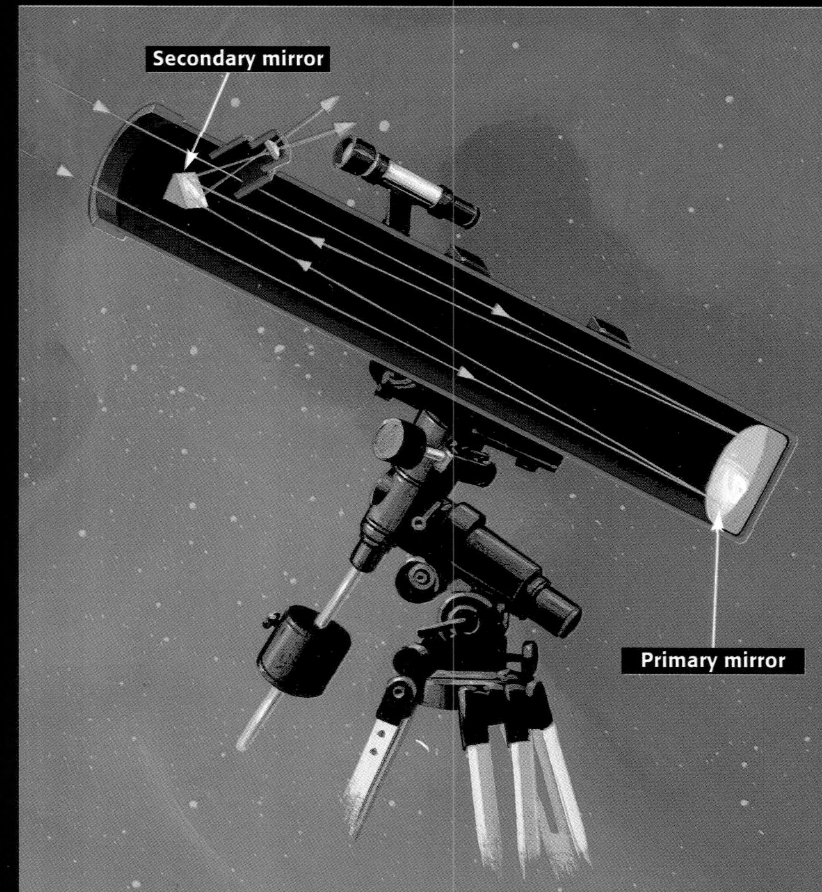

Optical principles of Newton telescopes.
[© *Ciel & Espace*/Manchu]

The Newton Telescope

This is the classic. One extremity of the tube, the one through which the light enters, is exposed to the outside air. The light rays strike the primary mirror at the end of the cylinder, reflect and converge toward the back. They're diverted 45° to the side by a perfectly flat secondary mirror. The light then exits through the side of the telescope where an eyepiece is installed to receive the image: the observer's gaze isn't directed toward the star, but inside the instrument. The secondary mirror is mounted upon a "spider," a support attached to the tube with metal prongs.

The spider partially blocks light from the primary mirror. Depending on the primary mirror's diameter, however, this obstructing effect should be relatively minor.

The Schmidt-Cassegrain

This telescope was the fruit of an original optical combination invented by the American Thomas Johnson, amateur astronomer and founder of the Celestron company. The primary mirror is spherical and pierced through its center. The front of the tube is sealed by a corrective lens, which diverts the light rays depending on where they strike it. A convex, secondary mirror is fixed upon this glass sheet. The light bounces back and forth between the two mirrors: it passes through the lens, lands on the latter, then "bounces back" and heads backwards. It exits through the aperture built into the primary mirror: the image, enlarged and corrected by additional optical devices, appears through the eyepiece. The marriage of the two basic approaches—refracting lenses and reflecting mirrors—has the major advantage of providing power, quality and reduced size and bulk.

Left: Newton telescope, Paralux Astro 115/900 on a manual equatorial mount.
[© SPJP-Paralux]
Opposite, above: Optical principles of Schmidt-Cassegrain telescopes.
[© *Ciel & Espace*/Manchu]

Secondary mirror

Lens cap

Primary mirror

Small, compact telescopes were invented to meet astronomers' need for short, light and easily transportable instruments. From 90 to 100 mm in diameter with focal lengths ranging from 500 to 1,250 mm, the highest-performance models can weigh up to eight pounds. They are often installed on a fork mount, equipped with drive motors and piloted from a control paddle (see sidebar, p. 192). This should make a fairly attractive first telescope, provided you purchase it through a reputable dealer and guarantee stability with a good tripod ("keep your feet on the ground and keep reaching for the stars!").

BIG, BIGGER, BIGGES

Amateur astronomers have l
114/900 (114 mm in diamete
length) or the 115/900 as begi
75 mm- and 90 mm-diamete
planets, stars, nebulae and gala
is accessible to the lucky ow
machines. Every brand offers o
the options offered by the s
upgrades, exchange of defe
important factors in choosing
motorized mount and stabl
115/900 is the first "all-in-one

Schmidt-Cassegrain telescope with Nexstar 8 Go To automatic slewing.
[© Médas]

Newton 46 cm diameter telescope, and NGT-18, on a horseshoe mount.

endless resources. It's upgradeable and can be adapted to different uses in return for a few changes in accessories. Beyond that, in the group of upgradeable instruments—from 130 to 160 mm in diameter—the observation site, telescope weight, your budget and the type of observations you hope to make will determine your purchase.

For the Initiated

Telescopes of 20 and 21 centimeters in diameter—Schmidt-Cassegrain and Newton—are the preferred instruments of dedicated amateurs. The former—widely available in specialty stores—are the most versatile. Their aperture ratio is nearly 10, which provides the high resolution necessary for planetary observation, while they collect enough light to remain effective for detecting faint stars. Knowing that the smaller an instrument's focal length, the more luminous it is, this telescope can be improved by grafting onto it a focal reducer, an optical device which reduces its aperture ratio to 5. The 21 cm are traditionally designed for observing the deep sky. They are most often installed at fixed locations and used by astrophotographers as luminous, super-telephoto lenses.

By convention, the so-called large-diameter telescopes are ones whose primary mirror exceeds 250 mm. These optical “monsters” collect a maximum of light and are best for observing nebulae and galaxies. Special mounts shaped like cubes, hoops or horseshoes complement these giant instruments. Surprisingly, their

Newton telescope available on the market the NTT 25, measures 635 mm in diameter for 3.15 meters focal length. Beyond that, the lover of observatory instruments will have to call upon private manufacturers willing to build mirrors one meter in diameter.

EYEPIECES

An eyepiece is a magnifying glass used to enlarge the image projected by the telescope lens or mirror. It's on this essential piece of equipment that we place our eye. As a general rule, purchase of your hardware will include two or three average-quality eyepieces of different magnifications. It's strongly advised that you complete this set with additional, higher-performance devices. In fact, eyepieces can transform and magnify an image when it passes through a quality lens. They offer greater—and necessary—suppleness to astronomical instruments: they magnify when needed, preserve high luminosity, adapt to turbulence, present perfect images and retain the depth of spatial vision. Last but not least, they can be fit on any instrument, past, present and future.

The Optical Formula

Luminosity, sharpness and final image contrast primarily depend on two factors: the quality of the lens glass—for perfect transmission of light—and the optical formula used. The latter, indicated on the eyepiece, varies based on the number of lenses involved and their arrangement. The simplest one, called Huygens and Ramsden, uses only lenses. The Kellner and AH (for Achromatic Huygenian) have three: they're better at correcting chromatic aberration, the tendency for light to scatter into rays of different colors. More complex and expensive orthoscopic and high-definition orthoscopic (also called multi-layer antireflective) eyepieces of the Erfle, Plössl and Super Plössl type contain four lenses.

Set of Plössl-type eyepieces for telescopes, from 6.3 to 40 mm focal length.

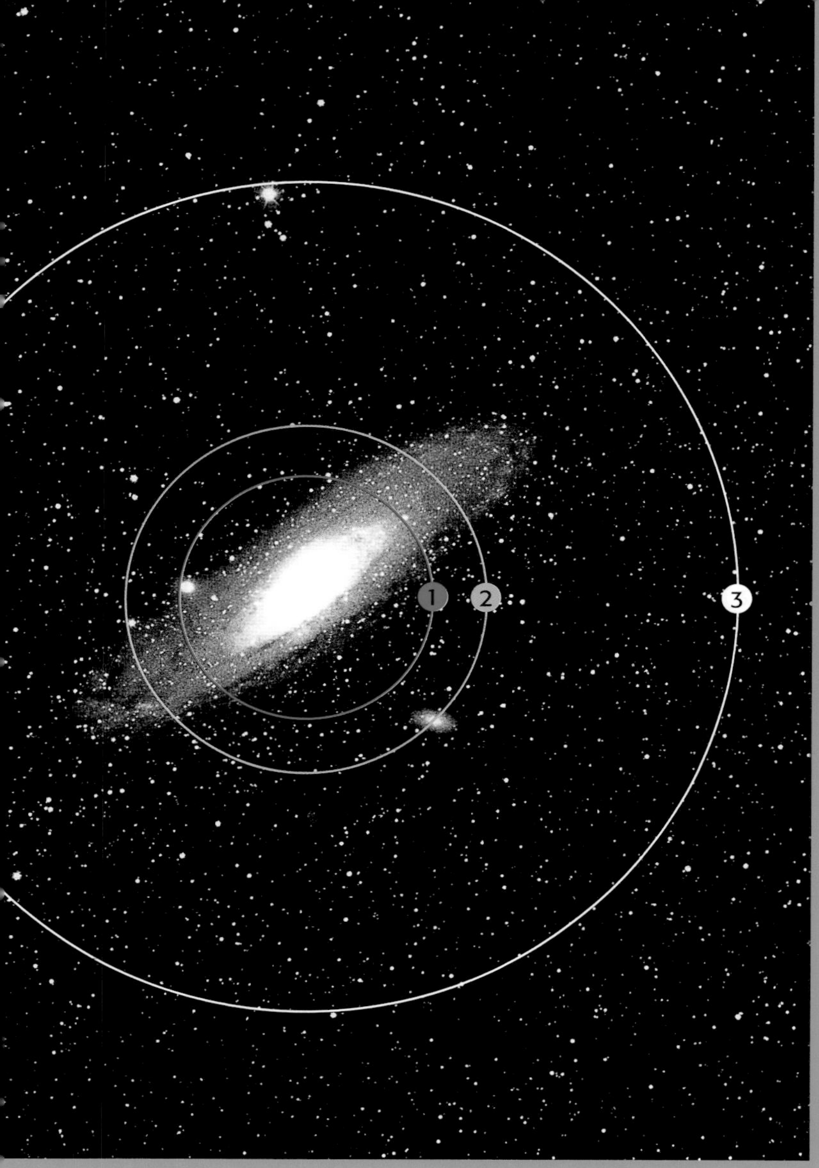

The instrument's field is the portion of sky accessible to the observer, which depends on the focal length of the instrument and the eyepiece used. Circle #1 represents the field of a Schmidt-Cassegrain 200 mm-diameter telescope, 2000 mm focal length, equipped with a 35 mm eyepiece. Circle #2 delimits the field of a 115/900 telescope equipped with an eyepiece of 40 mm. Circle #3 is the field seen through a pair of 10 x 80 binoculars.

[© *Ciel & Espace*/S. Anglaret]

Focal Length

The eyepiece, which contains a set of lenses, has a focal length too. The longer it is, the lesser its magnification. To calculate it, you need only divide the focal length of the instrument with that of the eyepiece. For example, a telescope with a 900 mm focal length, equipped with a 25 mm eyepiece, will enlarge the image 900/25 times, or thirty-six times. Eyepieces range from 2.5 to 40 mm in focal length. Which to use depends on the nature of the object observed, its luminosity and atmospheric turbulence. An eyepiece with a wide field and long focal length will reveal the subtleties of a nebula, while the details of a bright planet such as Jupiter can be seen on a calm night using heavy magnification.

Field

The field also depends on the eyepiece, the principal agent for image magnification, and must be taken into account when considering the portion of sky you'll be able to see. An eyepiece of 25 mm and 50° of field, for example, installed at the focal point of a 60/900 telescope, will provide a field of 1°38 for observation, or more than two and a half times the apparent diameter of the moon in the sky (0.5°).

Standards

The eyepiece diameters (called the "barrel" to designate the part introduced into the focuser) are classified by standards based on country of origin: Japan (24.5 mm) and American (31.75 and 50.8 mm). Adapter rings let you switch from one standard to another.

TRICKS OF THE OBSERVER

Stability is Key

In an ever-spinning world, comfort depends on stability. That is, on the tripod. A mounted telescope must be installed on an immobile tripod or column. Respecting this rule is essential. Exit, thus, tripods where vibrations and shaking ruin all efforts in pointing and following stars.

The World in Reverse

Careful: telescopes reverse the image of objects. Aim at a tree at the side of a road, and you'll see it upside down with the road on the other side. The world is reversed: high is low and right is where left should be. This problem, annoying for observing terrestrial objects, is partially corrected with the aid of a corrector. Provided with every telescope, it's placed upon the focuser and sets the image right-side-up. With the majority of beginner telescopes, for mechanical reasons, its use is required for observing the infinite. For the rest, such correctors are useless for viewing the sky, where up and down don't exist.

Instrumental Turbulence

Enemy number one for an astronomer is turbulence, both in the sky and in the instrument itself. Before using your telescope, it has to reach the right temperature. You need to set it up and let it sit at the observation site for a good hour beforehand. This is done so the telescope's own temperature adjusts to that of the surrounding air. Schmidt-Cassegrain telescopes, whose lenses seal the tubes, are less sensitive to instrumental turbulence. With the Newton, inside which air circulates freely, you must keep it away from all sources of heat, whether they be artificial or human.

Filters

Filters are used to improve the visual and photographic quality of images. They're divided into three families: solar, colored and interferential. The first are to be used with caution. Because of the power of the sun's rays, it's imperative that you limit the quantity of light entering the telescope. If possible, try to place a filter over the lens or front of the tube that reduces luminous intensity by 99.999%. After the total solar eclipse on 11 August 1999, an A4 format, safety filter film called "Astrosolar" became commercially available. With excellent optical qualities and ability to transmit light, this filter fits binoculars, cameras or astronomical instruments. In the higher range,

Set of colored filters designed to improve astronomical image contrast.

solar filters made of treated glass called full-aperture fit the majority of telescopes and come in all diameters. Colored filters reduce glare and accentuate the contrast of planetary surfaces. The interferentials, much more expensive, only let through electromagnetic rays corresponding to certain celestial objects. A UHC filter, for example, which, like colored filters, is fixed upon the end of the eyepiece closest to the focal point, lets you photograph hydrogen clouds of nebulae while blocking out the stray light of streetlamps. The price to pay is an overall reduction in luminosity, equivalent to the loss of one to two star magnitudes.

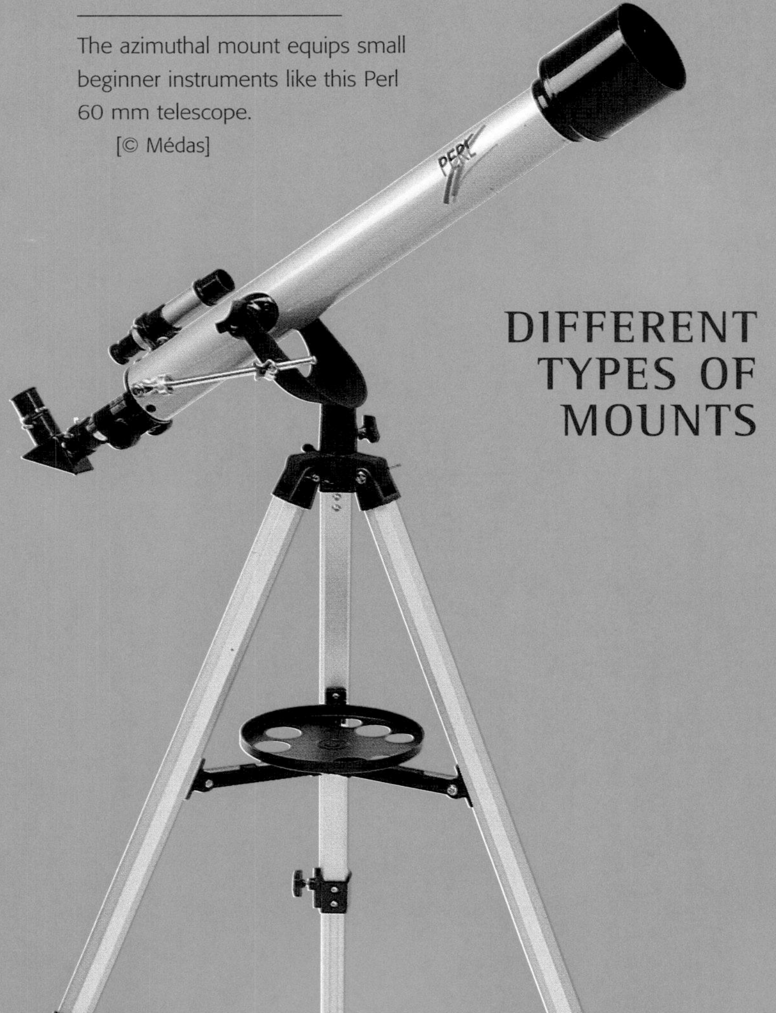

The azimuthal mount equips small beginner instruments like this Perl 60 mm telescope.
[© Médas]

DIFFERENT TYPES OF MOUNTS

A telescope's mount is the support on which the optical tube rests. They come mainly in two types: azimuthal or equatorial. Both are movable, so you can track the stars as they drift across the sky. Over the course of the night, the stars follow—in the east-west direction—arc-shaped trajectories perpendicular to the earth's axis. Their movement is apparent, in reality due to our own planet's rotation. To keep a star or planet in your field of vision, you need to compensate for its movement by turning your instrument in the same direction and at the same speed.

Azimuthal Mounts

These mounts are good for small, beginner instruments. Very simple to use, they need no initial adjustments. Azimuthal mounts move upon two axes: from top to bottom and left to right (azimuthally). Two stiff thumbwheels (brakes) halt its movement and two flexible joysticks help the observer pinpoint more exactly the desired object. It isn't ideal for following the movements of stars: you have to move it in two directions at the same time, both upward and to the right or left (depending whether you're looking southward or northward).

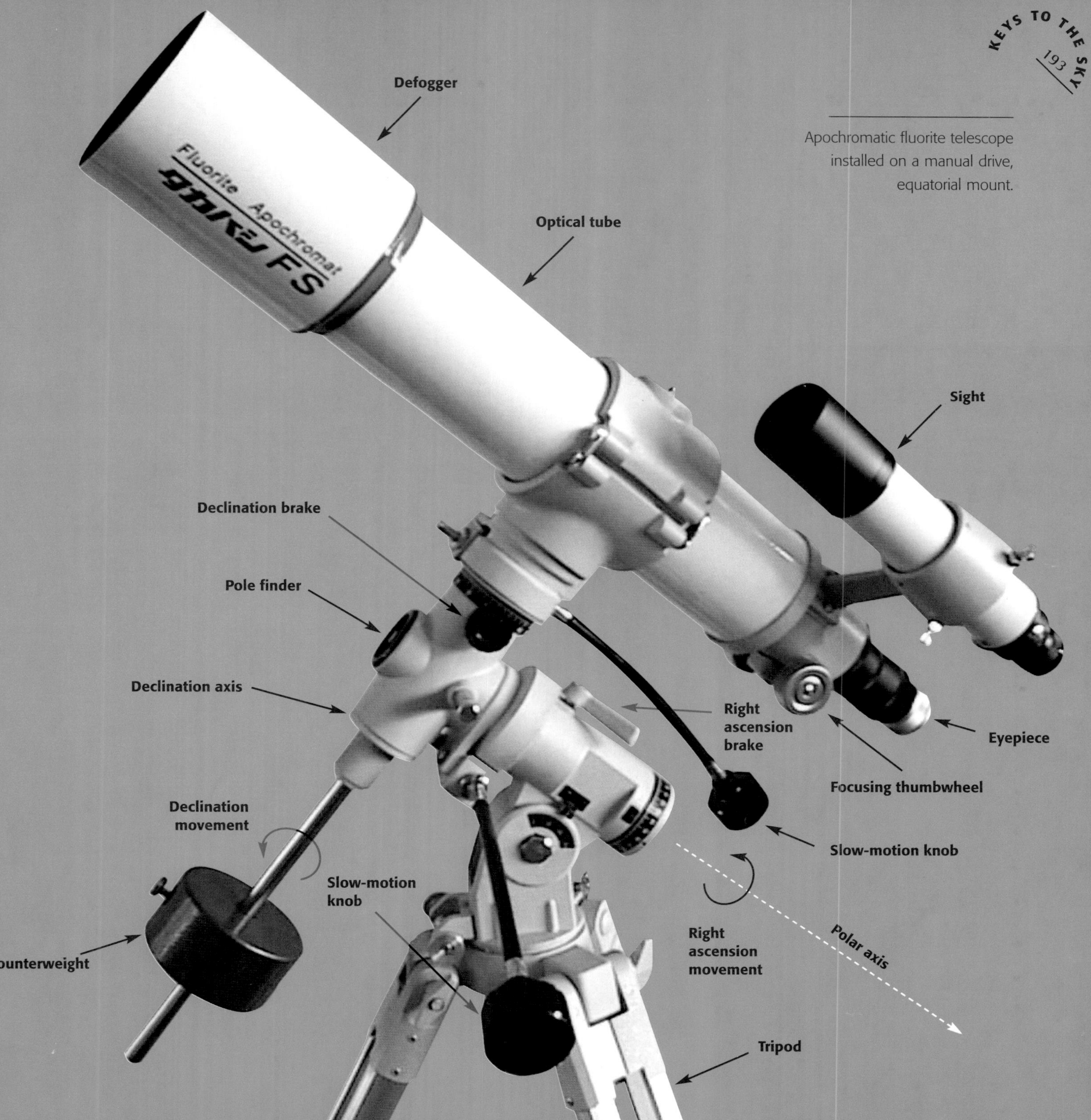

Apochromatic fluorite telescope installed on a manual drive, equatorial mount.

Equatorial Mounts

These mounts provide a tilted vertical axis in the direction of the celestial poles. They move parallel to the apparent trajectory of celestial bodies, which greatly facilitates tracking them. The position of a star is no longer terrestrial (with a height and azimuth you need to constantly correct over time) but celestial, with a right ascension and a declination. The former is measured in hours, the second in degrees. These coordinates, called equatorial, are universal and let you find the position of a star wherever you are on earth.

First off, an equatorial mount is more complicated to use than an azimuthal one. Its principal axis must be perfectly aligned with the celestial pole, while keeping a star in sight requires a certain mastery at manipulating movements in right ascension and declination. Today, manufacturers integrate a "pole finder" in the axis of the mount. This is an optical system composed of faintly lit concentric circles, which helps you find the pole directly and simplifies "positioning" to the extreme. Once the mount is aligned with the earth's axis of rotation, observers must find—all by themselves—a well-known, bright star, and place it at the center of the ocular field. They then find their equatorial coordinates in the ephemerides, according to which they align the pointer of the instrument's circle of declination so it's perfectly adjusted. This is called instrument pointing.

Autopilot

A large selection of equatorial mounts are commercially available. The so-called German mounts support telescopes of 115 to 150 mm, and weigh up to 11.2 lbs. Fork mounts are used for heavier, large-diameter instruments.

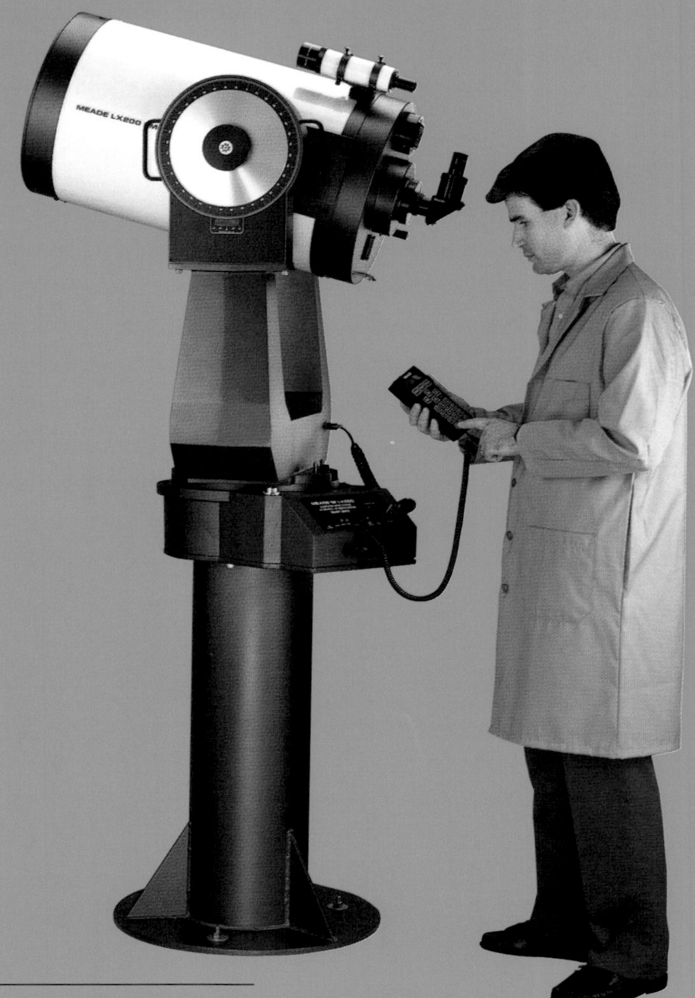

Meade Schmidt-Cassegrain telescope of 406 mm, on an altazimuthal mount, with automatic telescope pointing paddle [© SPJP-Paralux]

They're stable, easy to handle and feature electronic steering elements in their base. These two types of mounts are often equipped with tracking motors—for right ascension and declination. They resemble autopilots for airplanes or boats: they set the course for the observer and they guide the mount so the instrument stays constantly aimed at its target. Their mechanical qualities are determining factors for beginners in astrophotography or digital imaging. In these fields, exposure times are long, and you have to compensate for the movement of the earth's rotation perfectly so the stars remain point-like in your photos.

The automation of this process, tied to the development of computer science, contributed throughout the 1990s to the appearance of a new type of mount: the altazimuthal. This mount functions according to a vertical and horizontal axis (like the azimuthal mount) and is steered by a digital system that keeps the coordinates of the stars and major celestial objects in its memory. You zero in on a known star at nightfall—say, Betelgeuse. You then just type in its name to initialize the system to gain automatic access to every other star in the spangled vault.

A final type of mount, the Dobson (named after its inventor) is very popular, especially in the United States. It consists of a rudimentary azimuthal mount—a square, wooden box that pivots on a base—whose principal benefits are simplicity and cost. It's only used for visual observation and can support very large telescopes.

TELESCOPES

The Digital Revolution

Computers and telescopes have gotten married! For every model and diameter size, the mirror-chip couple have lined the shelves with novelties. It's now possible to buy a telescope and take a guided tour of the major celestial curiosities using automatic computer programs. The underlying principle is simple: a small computer hidden inside a large remote control is connected to the mount of the telescope (see sidebar, p. 192). It has a catalogue of thousands of stars, nebulae and galaxies in its memory. It can calculate the ephemerides of planets and quickly steer

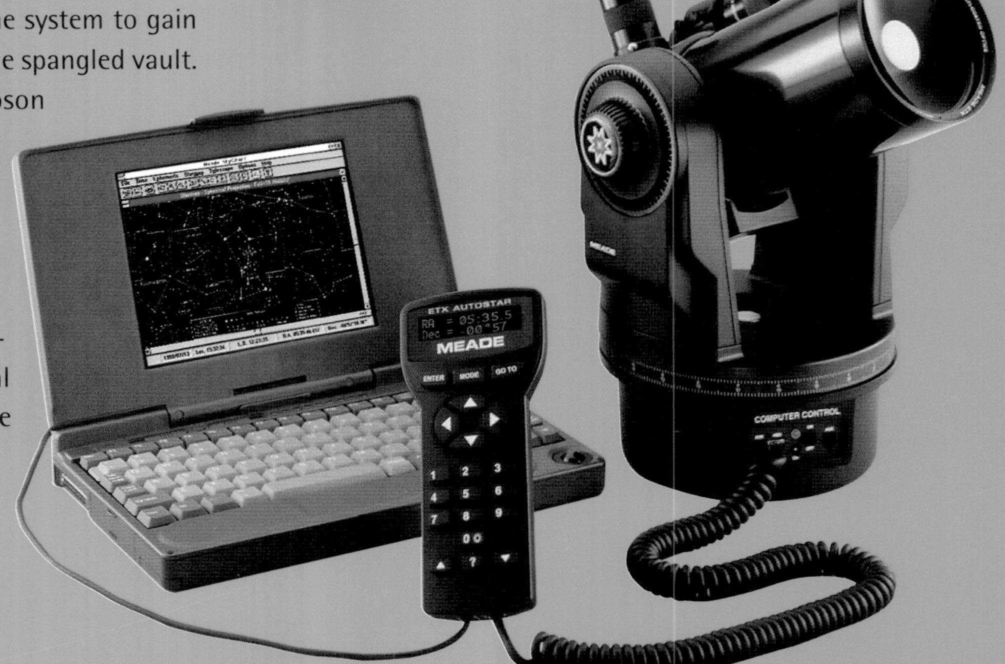

Computer, celestial cartography software, Autostar control paddle. With a simple click, the Meade ETX telescope points at the desired star.
[© SPJP-Paralux]

Schmidt-Cassegrain telescope, Meade LX 200, on an altazimuthal mount with a motorized drive.
[© SPJP-Paralux]

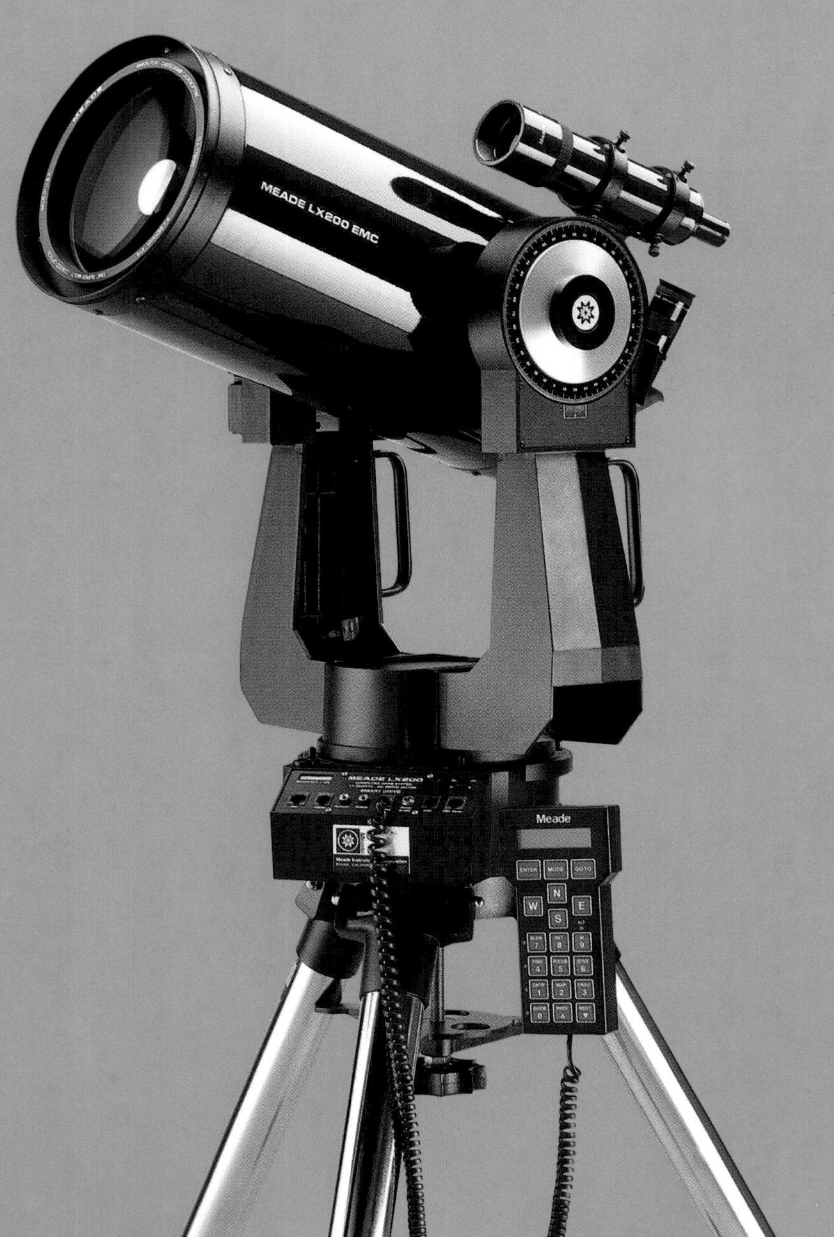

the drive motors to bring the desired object into view. You need only launch the program by specifying the date, hour and position of the site, and by picking out two known stars—stored in the computer—to confirm its positioning. It already knows everything that's up there . . . .

It's tempting to surf around the heavens by typing the name of a desired object on the control paddle (the "Go To" system) or by using the celestial catalogue available in its memory. Even better, specific consoles let you steer the telescope using a PC and preprogram an entire

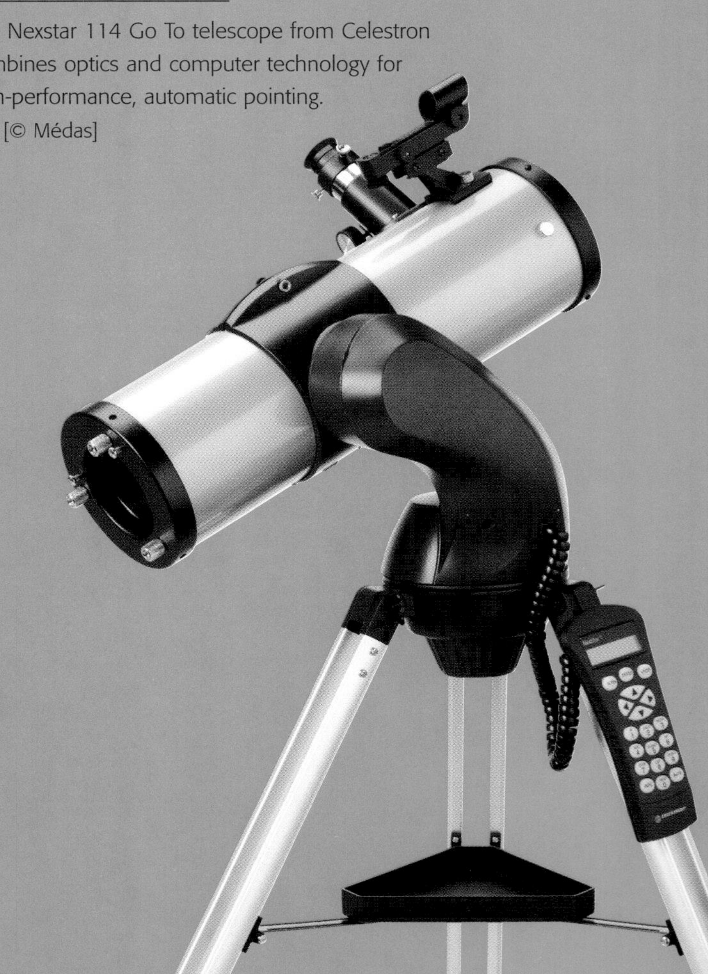

The Nexstar 114 Go To telescope from Celestron combines optics and computer technology for high-performance, automatic pointing.
[© Médas]

night of observation. Here, too, mount quality is essential (see sidebar, p. 192). The mount must be perfectly stable and equipped with tracking motors able to adapt their tracking speed to long exposure times.

These computers, with names such as Sky Sensor, Nexstar and Autostar, provide the amateur with instruments directly inspired by observatory telescopes. You no longer need to study the sky to unearth its curiosities: the digital revolution has taken hold of the universe. The advances have been remarkable. One of the primary telescope manufacturers—the American firm Celestron—mass-marketed a telescope equipped with a GPS (Global Positioning System). The satellite connection initializes the telescope's computer with the geographic coordinates of the observation site and calibrates to the exact second the positions of celestial objects. All that's left is to gaze . . .

How Much Does It Cost?

The world of compact telescopes opens its doors at a starting price of $300. The basic telescope, the famous Newton 115/900, is available for a budget ranging from $460 to $920. This price difference—a 1 to 2 ratio—is justified by the quality of the materials making up the optical tube,

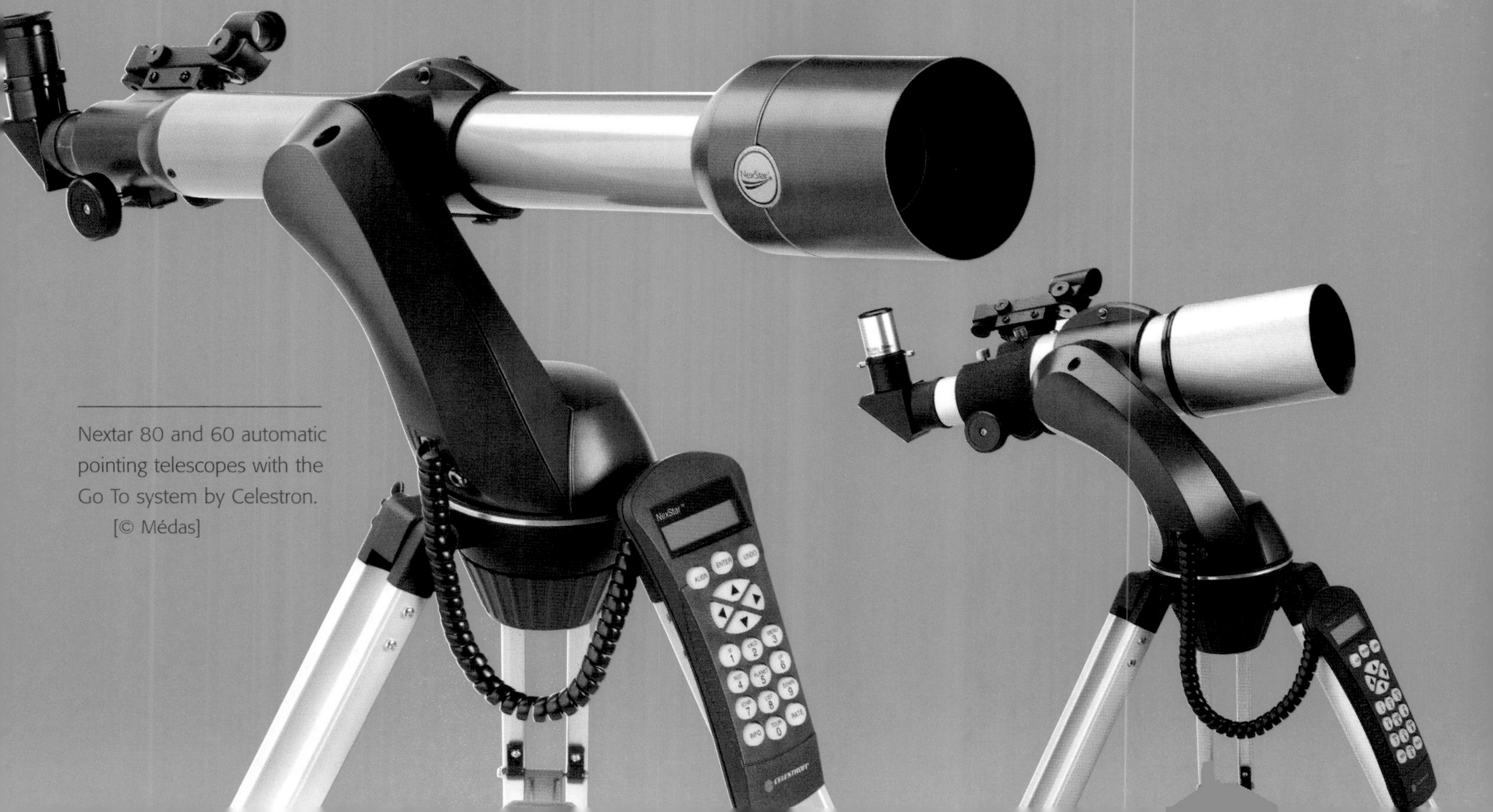

Nextar 80 and 60 automatic pointing telescopes with the Go To system by Celestron. [© Médas]

Newton telescope, on a fork mount, assembled by amateur astronomer Michel Peyrot.
[© *Ciel & Espace*/E. Perrin]

the popularity of the brand, the mechanical precision of the mount and possibility for upgrading—in sum, by a number of criteria you need to consider before making your decision. A good deal may lower the price to the detriment of quality: one more reason for paying close attention to the level of mechanical precision, the quality of the components, the stability of the mount and tripod, and the warranties and services offered by the store.

With larger diameters, from 120 to 160 millimeters, the rules are still the same: the best in optics and mechanics always come at a price! They range from $760 to $7,600, a tenfold difference. Excellent instruments cost between $1,150 and $1,830. The 203 mm Schmidt-

Cassegrain is reasonably priced today: from $1,760 to $5,260, depending on the type of mount and pointing systems you choose. The greater the diameter, the higher the price. $6,000 on average for a 280 mm-diameter telescope, $11,500 for a 355 mm diameter, and up to $38,110 for a monster 635 mm diameter—the price of a good automobile. An arsenal of three quality eyepieces will raise the bill anywhere from $230 to $550; a set of filters—solar, polarizing and Skylight—and other accessories will increase the burden another $300.

Building Your Telescope

Should you buy or build your telescope? For decades, this question has split the little world of amateur astronomy. The prices of commercial instruments, primarily from Japan, the United States and Germany—and thus sensitive to exchange rates—made acquiring a small 115/900 type instrument possible, but not a 21 cm telescope, reserved only for those with the deepest of pockets. A number of amateurs therefore chose to build their own instruments, which required long hours of labor and a certain talent for optics and mechanics. You begin by cutting and polishing a rough mirror to give it a perfect parabolic shape. Then, you have to construct the tube, focuser, and mount axis before installing a mechanic and electric drive system. From 150 to 200 hours of work for a total cost of around $1,000, labor not included . . . .

Commercially available astronomical telescopes are practically unlimited. So are their prices. A Meade 30 cm telescope.
[© SPJP-Paralux]

3

The Solar System

1. EPHEMERIDES 2. THE SUN AND THE MOON 3. OBSERVATION OF THE PLANETS

The universe? It's right next door! Want proof? An astronaut on the international Alpha station, revolving permanently around the earth, floats and lives in space 250 miles above the home planet—the average distance covered by the Eurostar between Paris and London. If a rocket lets you travel at the speed of light—approximately 186,000 miles per second—it would take you a bit more than a second to reach the moon and eight minutes to cook on the surface of the sun. You'd reach Mars in about twenty minutes, brush up against Pluto, the outermost planet in the solar system, in around six hours and discover Proxima Centauri, the nearest star, after a voyage of 4.23 years.

Observing the solar system is like "tending one's garden," or taking a careful look at one's surroundings. The moon and planets are the first and best destinations for an observer equipped with a telescope. Sunspots, a slender Venus crescent, a first quarter moon whose shadows outline the tormented relief of its seas and craters, a storm on Mars, the passage of a moon across the globe of Jupiter—there's always something going on in the sky. Of course, the movement of the celestial mechanism's many parts are often quite subtle and discreet. The seasonal events that give rhythm to the heavenly bodies circling our star don't make grand spectacles. But, sometimes, the passage of a new comet, an asteroid crossing the earth's orbit, or the promise of a meteor shower disturbs the monotonous hum and draws crowds around the telescopes. For some evening onlookers, this is the beginning of a long voyage toward the planets.

Twice a year, for an observer standing at the center of Paris's Champs-Elysées traffic circle, the sun sets directly inside the Arc de Triomphe. A spectacle you need to be in the right place at the right time to see.
[© *Ciel & Espace*/J.-P. Landragin]

1.

EPHEMERIDES

The program of such celestial spectacles is known in advance. It's published every year by the Institute of Celestial Mechanics in France and is available, in its entirety, in the astronomical ephemerides. By definition, ephemerides are tables that provide from day to day (or on regular dates) the positions of the stars and keep track of their changes. With the latest theories concerning the movements of the sun, moon and planets, astronomers in charge of calculating ephemerides have particularly effective tools of prediction at their disposal. Data can be found in official publications, such as the *American Ephemeris and Nautical Almanac* and the United Kingdom's *Nautical Almanac and Astronomical Ephemeris*, as well on the Bureau of Longitudes' Web site at www.bdl.fr/ephemeride_eng.html. From day to day, the ephemerides are picked up by the media—newspapers, radio and television—to announce sunrise and sunset times, the visibility of the moon and its phases and eclipses, and dates and hours of solstices and equinoxes. They're more detailed and illustrated in specialized reviews, which give advice for following events such as the conjunctions of planets and occultations and eclipses between planets and their moons. The ephemerides answer a simple question: which celestial bodies can we see tonight? Better, they announce the events marking the night to come. For amateur observers, reading them is edifying and lets them concoct a program for viewing specific occurrences.

DAY AND NIGHT

The first dancer in the ballet, the sun sets the key. The ephemerides begin by providing the hours of its rise, passage over the meridian and setting, and indicate the duration of the true solar day (the time taken by our star to complete its daily pass across the earth). In the Northern Hemisphere, the longest day is that of the summer solstice and the shortest that of the winter solstice. In France, times are generally provided for the Paris Observatory and, in England, for the one in Greenwich (located at the international source meridian, 0° longitude). You have to correct these times depending on your geographical location, empirically

Moon-Venus lakeside conjunction.

or by interpolating, arithmetically, the data of two known positions. Small distances can have large effects, though: in Strasbourg, the sun rises fifty minutes earlier than it does in Brest.

The duration of twilight is another piece of data the ephemerides provide. This information lets you know the true start of night. The tables distinguish the civil, nautical and astronomical twilights, whose respective durations vary with latitude. The first, civil, begins at sunset and ends at the moment when the center of the sun has dropped 6° below the horizon. The planets and stars of first magnitude appear and motorists begin turning on their headlights. The second, nautical, ends when the sun reaches 12° below the horizon. The horizon is still visible and the coastal lighthouses start to light up. The North Star and the Big Dipper can be made out. Night only fully arrives with the end of the astronomical twilight, when the sun has dived more than 18° below the horizon. Dawn, between the end of night and sunrise, is the exact symmetrical opposite of twilight.

The Moon's Agenda

The lunar ephemerides provide the day of the lunation (from one to twenty-nine) and its phase (from the new moon through last quarter), the hours of its rise, passage over the meridian and setting, the apogee and the perigee of its orbit (maximum and minimum distances of our satellite to earth), and its position in the sky at zero hours UT in right ascension, declination and parallax. This final information lets you locate the body with instruments and know the stellar field across which it's moving.

Ballet of the Galilean satellites—Io, Europa, Ganymede and Callisto—around Jupiter on 15 September 1997.
[© *Ciel & Espace*/A. Fujii]

September 15, 1997

12h01m31s UT

13h30m07s

13h50m00s

15h05m57s

15h09m36s

15h11m56s

15h50m26s

IV J I II III

Planetary Rendezvous

Reading astronomical ephemerides will give observers a fair idea of the conditions for viewing the planets. In the case of Mercury and Venus, the so-called inferior planets due to their relatively small distances to the sun, the most favorable moments for contemplating them correspond to their dates of greatest elongation, that is, when they're apparently furthest away from the daystar. The superior planets, from Mars to Pluto, are best visible during periods of oppositions, when they're located directly opposite the earth in relation to the sun.

Along with such information, data concerning the magnitude, apparent diameter and distance of the stars to earth simplify determining the best circumstances for observing using an instrument. You now need only figure out, by comparing these columns of data, the next long, moonless (and hopefully cloudless!) night when a majority of planets can be seen, high in the sky, at a moment when their trajectories bring them closest to earth. A dream!

The Dance of the Satellites

The precision of the equations of celestial mechanics takes on all its meaning when you read the ephemerides of the moons of Jupiter. The giant planet and its four principal satellites—Io, Europa, Ganymede and Callisto—discovered by Galileo in 1610, are a miniature, animated replica of the solar system itself. The tables of the "phenomena of the Galilean satellites of Jupiter" describe their ballet, punctuated with events such as their passage over the planet's disk (upon which they project their shadows), their eclipses (when they dive into Jupiter's shadow or that of the other moons) and occultations (when they disappear behind the planet's disk). Following the tables, contour lines aid in visualizing their positions in relation to Jupiter. When conditions are favorable, viewing this planetary mobile is impressive. Saturn, another target of choice for observers, is described based on the openness of its rings (this is maximal during the 2002–2003 period, and will shrink to a thin slice in 2009) and the elongation of its principal satellites: Titan (the brightest), Tethys, Dione and Rhea. Like for Jupiter, specific ephemerides describe the positions and the configuration of Saturn's other, lower magnitude companions.

The Little People of the Sky

After covering Uranus, Neptune and Pluto, whose apparent movements in the sky are very dim, the ephemerides provide the coordinates of the periodic comets, little planets and asteroids with magnitudes inferior to ten. Comets are bodies with known orbits that pass closest to the sun (at their perihelion) throughout the year (see sidebar, p. 216). The other types of bodies include large rocks like Ceres

The asteroid Toutatis, discovered by astronomer Alain Maury, is a geocruiser. It regularly brushes up against the earth's orbit.
[© *Ciel & Espace*/Cerag-J.-L. Heudier]

(577.8 miles in diameter), Pallas and Juno, followed by a crowd of names like Desiderata, Nina or Ursula suggesting the richness of their discoverers' internal voyages. In the sky, through a telescope, nothing apparently distinguishes them from stars; yet, from night to night, they move! Sometimes, when they pass in front of a distant sun, they block it out for a handful of seconds, revealing their own shapes and sizes. These occultations aren't visible from everywhere on earth, and the ephemerides specify on maps the zones where they're best admired. Naturally, they provide as well the date of the event and its maximum duration in a given location. When all the conditions are met, somewhere in the sky, an invisible hand seems to suddenly snuff out a star. A few seconds later, it lights up again. Magic!

Eclipses

Predicting eclipses and their general circumstances (beginning and end of their various phases) is another task of astronomers in charge of the ephemerides. The maps delimit the geographic zones for viewing total or partial solar and lunar eclipses, provide the conditions of visibility for a given location, as well as the time—in UT—of the phenomenon's various stages. Publications specializing in total solar eclipses (like the one maintained by Fred Espenak at NASA) are useful documents for observers: they describe how the solar corona (the sun's atmosphere) will look, enumerate the stars and planets closest to the sun at the moment of its disappearance, and even indicate the nebulosity and pluviometry of the zones in which all this can be seen. In any case, whatever happens, the event is sure to take place!

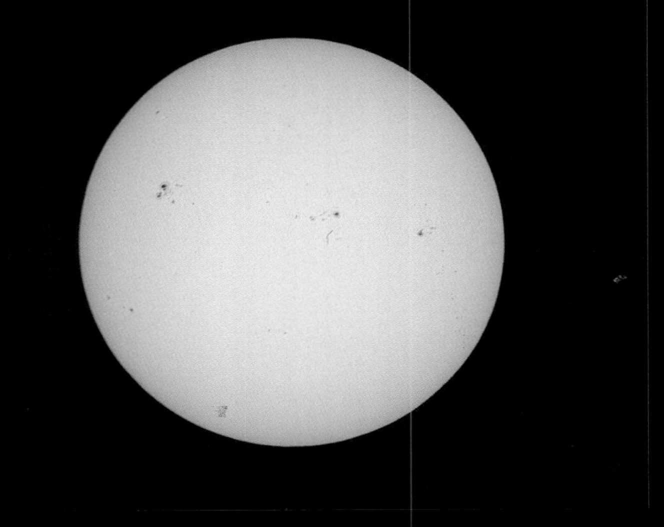

Spots on the solar disk testify to intense magnetic activity. They're sometimes visible to the naked eye
[© *Ciel & Espace*/P. Pelletier]

2.

THE SUN AND THE MOON

Two round faces, two big mugs as warm as they are inhospitable—such are the sun and moon. When the former disappears beneath the horizon, the latter pokes her head out and quickly imitates the daystar, chasing total darkness from the surrounding landscape. King of the day, queen of the night, the star and satellite have little in common besides being both attached to the celestial vault and, as far as such, the objects of our astronomical desires.

In comparison to many of his peers, the sun is a relatively calm star. Thanks to his placid temperament, life was able to develop on Earth over several billion years without winding up vaporized or frozen. However, the sun regularly shows signs of his evident magnetic irritability, the culminating point of a cycle that lasts eleven years. The quantity of visible spots on the sun's surface rises and, during this maxima period (year 2001 marked the summit of cycle #23), the solar acne is concentrated toward the zones nearest the equator. Eruptions multiply. The lucky

THE SOLAR CYCLE

owners of coronagraphs or interferential filters can track the evolution of the protuberances that accompany these celestial bursts of anger. On Earth, a gust of particles stirred up by these magnetic storms results in the polar auroras, where the sky begins to resemble luminescent curtains rippled by an invisible wind. This phenomenon occurs when a flow of charged particles ejected by the sun crosses the poles of the earth's magnetic shield. During the most intense of these storms, the auroras can be seen at the latitude of Paris, illuminating the sky like a tremendous bonfire.

OBSERVING THE SUN

Why wait until night to raise your eyes to the sky? The most brilliant of all the stars, the closest one, too, is the sun, and it's natural to want to point your telescope there. Provided you respect certain rules for contemplating it safely, the sight of its tumultuous bubbling is amazing. Irregularities shaped like grains of rice—they resemble bubbles breaking on the surface of a pot of boiling water—envelop the visible "surface" of the sphere of gas in a granular skin. In places, the sun's marbled with black spots sometimes large enough to be seen with the naked eye, were the sun not so blinding. These occurrences, the most complex of which come in groups, are bordered by a grayish penumbra, rimmed with thin arabesques. The spots are framed in a lighter border, faculae, when they appear on the edge of the solar disk. Finally, and provided you have a coronagraph or special filters at your disposal, it's possible to see flames leap off the solar edge and soar off into the surrounding space. These protuberances, which attest to the violence of the eruptions, can reach in several hours heights greater than the earth-moon distance and then . . . disappear!

Watch Your Eyes!

Observing the sun is dangerous. Too much light and heat are enemies of astronomical instruments designed to amplify the power of light rays using lenses and mirrors. You need only remember how quickly a magnifying glass sets fire to twigs or paper to comprehend the importance of respecting safety rules aimed at protecting you from irreversible ocular lesions. Looking directly at the sun is out of the question, and it's a good idea to place a hermetic seal over the

instrument's viewfinder to keep the tiny magnifying glass from accidentally burning your eye. To aim the instrument toward the daystar, rely on the shadow of your telescope tube projected on the ground. When it's at its shortest, that means it's lined up with the earth-sun axis. Next, it's imperative you equip yourself with filters to reduce the quantity of visible, ultraviolet or infrared light entering the telescope (see chapter 2). If this isn't possible, project the image onto a white screen held perpendicular to the eyepiece. Be careful not to use eyepieces with lenses whose glue will melt under the intense heat.

A GUIDED TOUR OF THE MOON

If the moon didn't exist, the face of the world would be completely different! We wouldn't have tides, wouldn't have eclipses, the earth would likely be unstable on its axis of rotation and our sky would be stripped of the most familiar of globes. The disaster would be total. Goodbye moonlight serenades, crescent

THE CORONAGRAPH

It was a French astronomer, Bernard Lyot, who invented this clever instrument in 1930. It lets you see eruptions around the sun's edges and, when the sky is very clear, the shape of its atmosphere, that is, its corona. The principle is simple. During a total solar eclipse, the lunar disk masks the daystar, and the red bursts of its protuberances, normally drowned out in the stream of light, become visible. At this time the corona, as bright as a full moon, reveals magnetic spikes and loops that shoot out far into space. With the coronagraph, Lyot created an artificial eclipse. Inside a telescope, he carefully placed a disk or opaque cone that obscured the star and, with the help of a filter, caused the sun's flames to appear!

Today, coronagraphs are installed in space to track solar eruptions or research faintly luminous objects around bright stars, such as exoplanets. For amateurs, however, the appearance of a new generation of "H Alpha interferential" filters has rendered their use obsolete. For around $1,000, these filters can be adapted to any instrument (with an aperture ratio equal or superior to 20) and offer a very original vision of our star's disk: a bubbling saucepan!

Eruption of solar protuberances observed using a coronagraph. [© *Ciel & Espace*/APB]

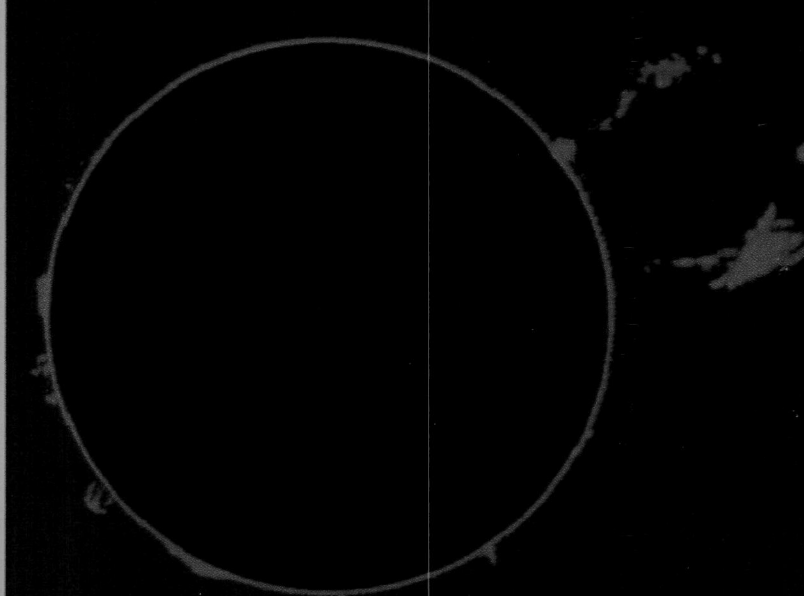

shows at twilight and dawn, cinematic monsters, werewolves baying at the full moon . . . and the favorite object of astronomical spectacles.

A Little Selenic Geography

Using just your eyes, and then a pair of binoculars for appreciating a 3D image of the globe, you'll first see the seas appear (in reality, plains covered in basalt),

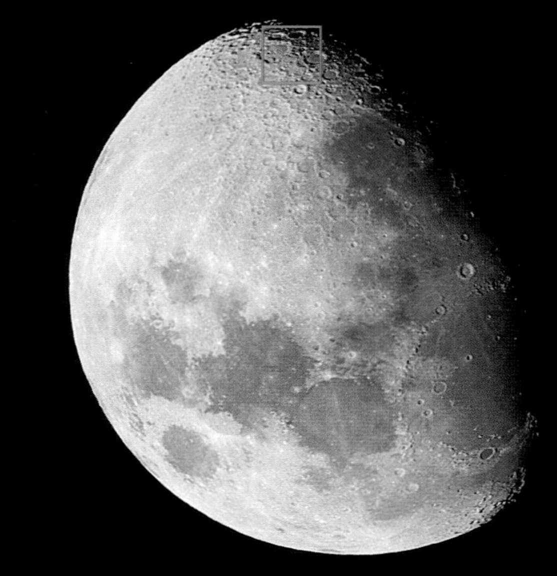

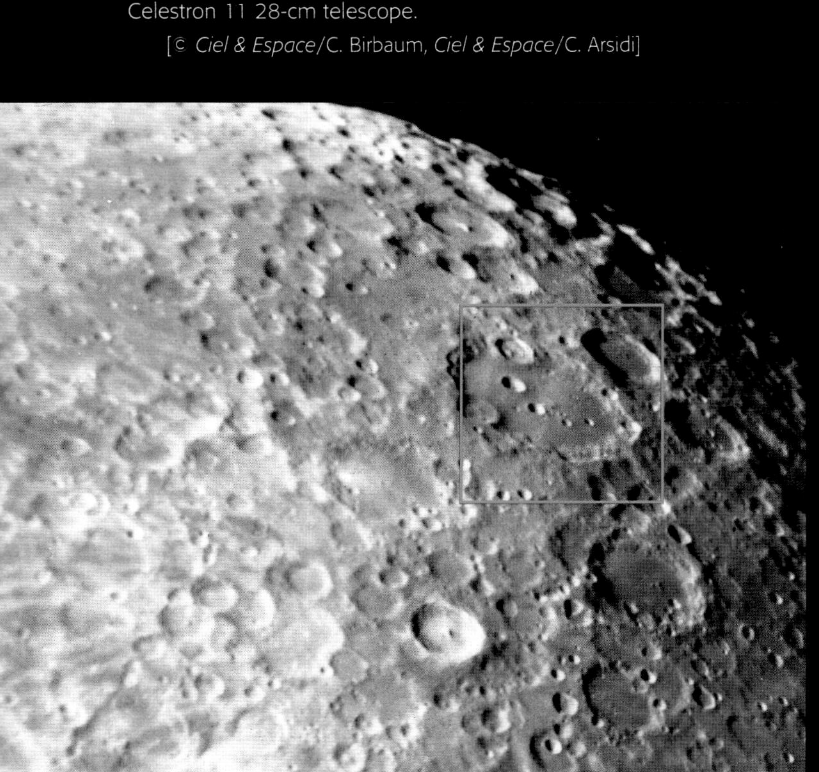

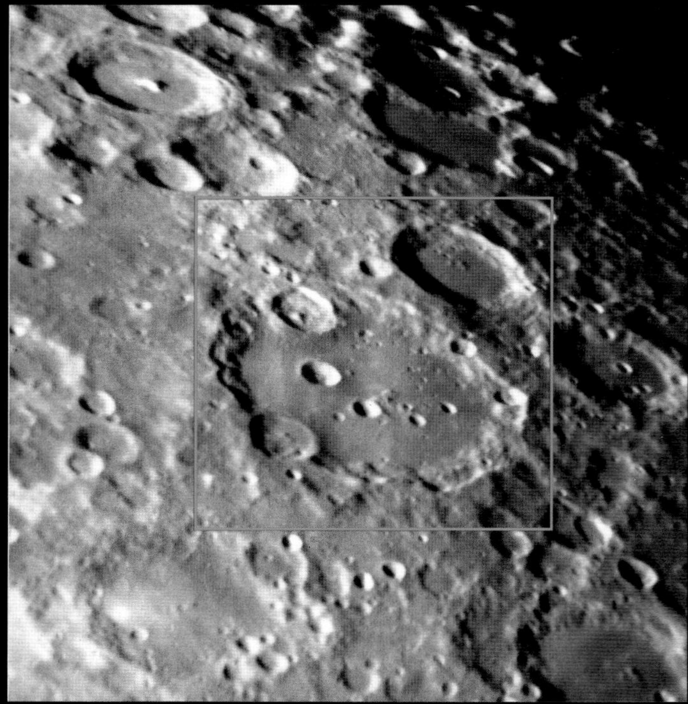

On the southern edge of the moon, the huge crater Clavius—140 miles in diameter—is a vast plain surrounded by very steep ramparts. Its rounded bottom is riddled with more recent craters. Global photo from a telephoto lens, close-up using a fluorite telescope of 90 mm, then Celestron 11 28-cm telescope.

then the continents—uneven, dusty zones riddled with craters. Their shades are different: the former are grayer and darker, the latter are brighter. The names of the seas—assigned in the seventeenth century—split the moon in two. Good weather and calmness reign in the first quarter thanks to the Seas of Fecundity, Serenity, Tranquility, Nectar, and the like. Then, atmospheric misfortune and catastrophe dominate in the last quarter with the immense Ocean of Storms and Seas of Showers, Cold, Clouds and Moisture. Some of them have very clear contours, others possess irregular shapes that branch out into structures called gulfs, lakes, swamps, highlands and plains.

With a telescope, visual exploration of the continents reveals innumerable scars born out of a history stretching back more than four billion years. Chains of eroded mountains, stripped of peaks and needles, border the seas and emerge out of the most damaged zones (near the South Pole, Mount Leibnitz rises 5,095 feet). Grooves, folds, faults and valleys are the visible consequences of fractures of the lunar surface due to earlier volcanism and blows dealt by violent meteoritic collisions. Millions of craters dot the Selenic surface (the greatest—Bailly—measures 178.3 miles in diameter, the deepest—Newton—reaches 5,280 feet). Cirques with steep walls and a central pinnacle, craters half erased or drowned in lava, chains whose alignments suggest multiple bounces or the breaking up of meteoritic bodies, intermingling craters, walled plains,

Remarkable conjunction between the moon and Jupiter, observed with a Celestron 8 telescope on 22 June 1983.

[© *Ciel & Espace*/C. Arsidi]

radiant systems whose bright streaks betray youth, all transform the moon into a cemetery of past disasters.

The Landscapes of a Young Moon

The complete cycle of lunar phases over 29 days, 12 hours and 44 minutes—called synodic revolution—lets us enjoy the different fractions illuminated by the sun from one night to the next. From 14 to 20 hours after the invisible new moon—the period of conjunction with the sun—observers will see a comma of the "young moon," thin as a silver thread, tracing its timid smile through the glimmers of sunset. Starting from the second day, it's easy to pick out the Sea of Crises, in the upper right-hand side, trapped between the limbo (the lunar edge) and the terminator, the line of demarcation between the lit part and the dark part. Here's where the most beautiful spectacles take place. The sunlight grazes the surface there, like light filtering underneath a door in a dark room, revealing every little detail as clearly as specks of dust on the floor. The first days of the lunar cycle are times of great shadows and—with the help of low turbulence—exceptional contrasts. The dark part of the Selenic globe shines feebly: this is the ashen light. The moon is reflecting the glimmers of Earth, lit up by the sun!

Games of Shadow in the First Quarter

On the seventh day, the first quarter is the ideal period for watching ephemeral details of the lunar relief come and go. It's best to begin your observations before nightfall to benefit from the low light: you'll be able to track the sunrise across the lunar peaks and craters. Systematic sweeping of the terminator, at high magnification, lets you stroll along the relief: from the north—where you'll find the crater Plato and the Valley of the Alps, near the Sea of Showers—to the south occupied by craters Tycho, Clavius and Moretus, and pass through those of Ptolemy, Alphonse and Arzachel.

TRANSITORY PHENOMENA

The moon is a dead body. On many occasions, however, observers (such as the Apollo 11 astronauts) have claimed to witness fog, glimmers, colored spots and lightning on its surface. These curious manifestations, which seem to occur most often at the edge of the seas and beside the moon's many grooves, have been labeled "transitory lunar phenomena." They were studied by NASA during the 1960s (and are still researched today by small groups of amateur astronomers), and their source is widely debated. The most common hypotheses invoke optical illusions, collisions between the moon and meteorites, or outgassing from the lunar subsoil resulting in little dust clouds briefly lit by sunlight.

In mid-November 1999, lightning was recorded on videotape by amateurs. It came from the dark regions and was filmed at the moment our system was crossing the swarm of shooting stars of the Leonids, a zone of dust caused by the outgassing of the comet Tempel-Tuttle.

The shadows accentuate the evidence of innumerable accidents that have struck the lunar surface: rocky pinnacles, grooves, dissolved ramparts and differences in levels over gentle slopes. Very careful focusing, performed away from any wind, makes all the difference to patient observers who know how to bide their time: when the air stops dancing, the moon will reveal herself.

The Craters of the Full Moon

With the waxing gibbous moon (when the lit regions are greater than the darkened ones), the lunar shadows shorten and grazing light reveals new regions at the terminator: the Sea of Moisture, the Ocean of Storms and, at the edge of the Sea of Showers, the Gulf of Iris, an apex of which forms the famous "woman's head." Named "Heraclides Promontory," this formation takes, around the tenth day, the appearance of a feminine profile partially concealing her long hair. Nearby, at the foot of the Jura mountains, careful observers will observe a "face of rock," the fleeting apparition of a human face whose traits are frozen for eternity in the dust.

The night of the full moon is, in principle, an evening of rest for astronomers. Fourteen days and twenty-eight hours after the new moon, or the beginning of the lunation, it's lit directly by the sun and nothing's visible in the sky beside our satellite and several bright planets. On this night, there are practically no shadows to provide contrast to the lunar relief. On the other hand, the direct spotlight illuminates the moon's younger materials, bright traces that attest to recent meteoritic impacts. The crater Tycho is a perfect example. More than 100 million years ago, a giant meteorite (around 6.2 miles in diameter) crashed into the southern portion of the moon. Beneath the terrible shock, material and dust were ejected a great distance. Silence then fell again across this new 52.8-mile-diameter crater, lost amidst the crowd of its peers. Its anonymity comes to an end when the sun is at its zenith. As if struck with a magic wand, Tycho lights up and becomes the center of a vast, radiant pattern across the moon. Starting from the point of impact, sumptuous stripes of bright powder spread out across the Selenic globe for over 1,240 miles. With a gray neutral filter placed in the eyepiece or, even better, a red filter which lets you emphasize the contrast of the stripes, Tycho's vast radial structures, along with those of Kepler and Copernicus, attest to the violence of the impacts that struck the bodies of our solar system.

The Show's Over

Shadow overtakes the eastern edge of the moon when it's in its descendant phase. It becomes gibbous again and visible later and later in the evening; its lit portion is rapidly nibbled away by darkness. On the twenty-eighth day, the last quarter rises in the middle of the night. Here again, the sun's grazing light lets you appreciate the undulations at the edges of the seas, the large craters' dusty floors, and the tangling of the formations at the South Pole, where astronomers suspect that water—brought by the fall of comets—mixed into the lunar surface. The final crescent rises several hours before the sun and, after twenty-nine days traveling around the earth, the moon takes her bow by melting into the glimmers of dawn.

3.

OBSERVATION OF THE PLANETS

"You must not expect to see at sight. Seeing is in some respects an art which must be learned." This wise advice by English astronomer William Herschel still holds true. Don't be fooled: observation of the planets is a difficult art. Between the photographs published in the magazines—often taken by automatic probes or spatial telescopes—and direct vision in the eyepiece of a telescope of a half-hazy confetti quivering like a cork in water, the distance is . . . astronomical! In reality, these two types of images are complementary. The former belongs to the history of space exploration and science. The latter belongs to the present, to the pleasure and talent of the observer. From day to day, the planets change in appearance, and numerous signs attest to their seasonal transformations. The distance of a body to the earth but also the quality of the observation site, atmospheric turbulence and the precision of your instrument are the rules by which you have to play the game. On a beautiful, calm night, the image of Saturn's colored globe in a telescope, intensified by the shadows of its rings as they cast themselves across the planet's atmosphere, is no less impressive than a photograph taken by the American Voyager probe.

The crescent of the planet Venus. Celestron 8 F/80 with a W 80 A filter. [© *Ciel & Espace*/E. Beaudoin]

VENUS AND MERCURY, THE SUN'S NEIGHBORS

Like the moon, Venus and Mercury present a cycle of phases. These planets are at first invisible when they slip between us and the sun (inferior conjunction), then present a thin crescent as they move away, show a quarter at the elongation (the largest angular distance between the sun and the planet), and finally pass into superior conjunction, behind the daystar, before reappearing in the gibbous phase. The smaller of the two, Mercury, never moves more than twenty-three degrees away from the sun and often remains drowned in its light. It's best visible during certain elongations, the most favorable of which, in the Northern Hemisphere, occur in the evening in early May and in the morning in October. In contrast, Venus, known also as the Shepherd's Star, can shine with all its fire (magnitude –4) and remain visible long into the night. Its maximal elongations exceed forty-five degrees and its thin crescents, during inferior conjunctions, are a real curiosity. With a telescope magnifying forty times, it looks like the moon seen by the naked eye. As with Mercury, no detail of Venus's surface can be seen. On several very rare occasions, the sun, the earth, Mercury or Venus are perfectly aligned. It's the moment to observe a "transit," the passage of a planet over the solar disk. The next transits of Mercury will take place on 5 May 2003 and on 6 November 2006; those of Venus, much rarer, are expected on 8 June 2004 and 6 June 2012. The event is important because Venus's last transit goes all the way back to 1882!

THE MARTIAN ORANGE

With Mars, the best conditions for observation come together during oppositions, when the sun, the earth and the red planet are aligned. Due to the very strong eccentricity of its obit, Mars's distance to the earth can double between its shortest and longest points—from 34.8 to 62.1 million miles. Every two years, it's possible to make out the white spot of its polar skullcap and identify, starting from several dark zones, the large formations that mark its surface: the grayish triangle of Syrtis Major, a craterized volcanic plain, for example, or Mare Acidalium whose expanse caused earlier observers to believe there existed a sea on Mars. With heavy magnification and a good dose of patience (and a lucky break in turbulence), a 100 mm or 150 mm telescope

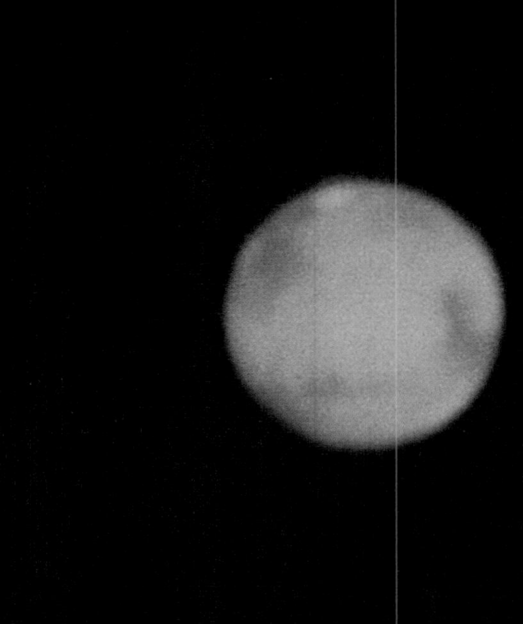

The planet Mars seen through a telescope.
[© *Ciel & Espace*]

provides access to several details of the planet's orange globe. But only an excellent 150-mm-diameter refracting telescope or a reflecting telescope of more than 200 mm will let you track the progress of dust storms, the regression of the polar ice caps or the birth of clouds and mist above Mars's deserts and volcanic summits. The use of colored filters, which increase the contrast, is indispensable for observing seasonal phenomena.

Mars will be in opposition in August 2003, the best occasion of the decade to observe the mythic planet up close. Still, at 150 times the earth-moon distance . . .

The giant planet Jupiter with the passage—at the bottom—of its moon Ganymede.
[© *Ciel & Espace*/C. Ichkanian]

The familiar universe of the telluric planets made up of solid materials like our planet and on which we can easily imagine standing is turned upside-down by Jupiter. The largest lodger in the solar system—two and a half times the mass of all the others put together—is a beautiful, gaseous globe, sharply flattened by its speed of rotation (it makes a full revolution around its axis in nine hours, fifty-five minutes). A pair of binoculars reveals a small, bright and uniform disk, along with the four moons discovered by Galileo in 1610 (Io, Europa, Ganymede and Callisto). Through a 60 mm telescope and a magnification of 100 appear two dark, cloudy strips across Jupiter's equator, as well as bright zones inside which circulate white clouds. The famous Great Red Spot—a gigantic anticyclone whose size nears twice that of the earth—is visible with a telescope of 115 mm about twenty degrees south of the giant's equator. Since its discovery in 1664, its colors have constantly varied: it can resemble an eye of black butter or almost become invisible. During favorable oppositions—at four times the earth-sun distance—the details accessible through a 100 mm to 150 mm refracting telescope or a 200 mm to 300 mm reflecting telescope are impressive. The use of colored filters (placed inside the eyepieces) allows for tracking the movements of the great Jovian formations. Like a river stirred up with currents and eddies, the turbulent upper atmosphere accompanies the rotation of the planet by stirring up its clouds, spots and plumes. What a spectacle!

Four of Jupiter's sixteen known moons are bright and easy to observe. Since they revolve around the planet in very short time periods, the best way of identifying them is to note their positions in relation to Jupiter's disk, then refer to the graphs published in the ephemerides (see above). Observation of their movements in front of (passages), in the shadow of (eclipses) or behind (occulations) the Jovian disk illustrates in real time the precision of the laws of celestial mechanics. Following the passage of a Galilean moon—then its shadow—in front of and over Jupiter's globe through a high-magnification telescope is a magic and silent spectacle. Every six years, when the earth and the sun cross the plane of their orbits, two Galilean moons can mutually occult or eclipse each other. This theater of shadows is a meeting point for the initiated . . .

SATURN AND ITS RINGS

Even further, at ten times the earth-sun distance, Saturn is the most fascinating member of the solar system. Simple, yellowish and point-like through a pair of binoculars, the planet looks through the eyepiece of a 60 mm telescope like a small, off-white pearl, noticeably flattened and crowned by a system of rings. It's difficult to make out its details: during its best oppositions with the earth, the entire system—globe and rings—reaches the apparent diameter of Jupiter. With a 100 mm refracting telescope or a 200 mm reflecting one, high magnification lets you make out dark and bright bands—like on Jupiter—evidence of cloud movement in the upper atmosphere. But the real spectacle is elsewhere. Three distinct structures make up the

COMETS

Questioning Bill about the nature of the hazy spot he just snuck up on through his friend's telescope, Thomas Bopp learned he was the lucky and innocent discoverer of a comet. At the same moment and several hundred miles away, in the mountains of New Mexico, Alan Hale, an unemployed professional astronomer, noticed thanks to a homemade telescope the movement of a suspicious nebulosity. According to celestial maps, there shouldn't have been anything but stars in that spot. Hale-Bopp, the most recent bright comet observed at our latitudes (in 1997) is, with the great Hyakutake (in 1996) a typical illustration of the way comets are discovered: thanks to a healthy dose of luck and a smidgen of chance, often complemented by meticulous and obstinate research.

Amateur astronomers are, besides space probes and computer programs seeking asteroids (rocks, the majority of which are in orbit between Mars and Jupiter), the most frequent discoverers of these hairy bodies. When they approach the sun, these "dirty snowballs" heat up, and the ice they contain is released in the form of gas and dust. To discover a comet, an amateur must pinpoint, first, a new nebulosity against the starry background, note its movement on a sky map and have the discovery confirmed by the Central Bureau of Astronomical Telegrams. At the end of this process, the new body will bear the name of its lucky inventor.

Beautiful comets visible to the naked eye are rare. In a perfectly black sky, their tails of gas and dust stretch for hundreds of millions of miles. They disperse into space the materials they contain. In a telescope, the head of a comet is the seat of heavy activity when it passes closest to the sun and begins to melt. It changes shape and brightness, its center starts erupting and sometimes breaks into pieces. When the core (its heart) is active and emits large quantities of dust in continuous jets, these streaks reflect the sunlight and let off yellowish curls, rapidly swept away by the solar wind.

With large-diameter binoculars, whose field and luminosity are sufficient for observing both the comet's head and tail, the spectacle is amazing. The effect of binocular vision reinforces the impression of relief, of three dimensions created by the transparency of cometary nebulosities.

The last great comet, Hale-Bopp, during its passage near the earth in April 1997.
[© *Ciel & Espace*/D. Pankowski]

Saturn, the ringed planet, observed with an Astrophysics L 155 F94 EDFS telescope.
[© *Ciel & Espace*/C. Ichkanian]

system of rings. Of different colors and brightness, each ring is separated by two black strips of empty space: Cassini Division—the most visible one, like a line of India ink—and the Lyot Division. The 3D impression is striking: the shadow of the rings is cast across the planet, which eclipses part of the ringed disk. On the equatorial plane, Titan, the largest of Saturn's satellites (one and a half times the size of our moon) is only a luminous point, indistinct from a star.

Observable phenomena on Saturn are almost nonexistent. The only event of importance is the passage of the earth, every fifteen years, through the plane of the planet's rings. At this moment, their profile, whose thickness compared to the diameter of the planet is equivalent to a sheet of cigarette paper placed before an elephant, seems to totally disappear. Since the most recent disappearance, in 1995, the tilt of the system has been slowly increasing and will offer its maximal openness during the 2002–2003 period. Then the process will reverse itself and, in 2009, the rings will be seen at a cross-section for a period of several weeks—on the astronomical scale, the blink of an eye.

AT THE EDGE OF THE SOLAR SYSTEM

Uranus, Neptune and Pluto, the three final planets of the solar system, are inaccessible to observation. At best, Uranus, which isn't visible to the naked eye, through the eyepiece of a 200 mm telescope and enlarged two hundred times looks like a distant, greenish, hazy disk, totally stripped of detail. Neptune's movement can be tracked, from one night to the next, against the celestial vault's background of stars. As for Pluto, discovered in 1930 by American amateur astronomer Clyde Tombaugh, only the experienced astrophotographer will be able to enjoy piecing together over several months its path across the sky. A good tracker of unidentified celestial orbits knows that everything that moves belongs to the solar system!

4
Deep Space

1. THE STARS 2. THE NEBULAE 3. THE GALAXIES

There exist more stars in the sky than grains of sand on the surface of the earth! This concept is astronomically huge, as disconcerting as it is inconceivable. Reality isn't any less bewildering. On a beautiful summer night, a glance across the starry vault is enough to make out the white and luminescent trace of the Milky Way, first witness to this plurality of worlds. This hazy strip is the disk of our galaxy, viewed at a cross-section: a group of 100 to 300 billion stars dragged along in the same movement of the giant, cosmic wheel.

From north to south, the Milky Way crosses the constellations and cuts the sky in two, before forming an arch whose pillars sink beneath the horizon. The darker the night is, the clearer the contours of its milky shape appear. With a pair of low-magnification binoculars large, hazy spots will appear, clouds lit from the inside by innumerable groups of stars. The extreme richness of this multitude of stars reveals itself at the first glance cast through a telescope. The vision is subtle, enchanting. Initiates, familiar with what's happening beyond the solar system, qualify these plunges into space-time as observing the "deep sky." It provides access to the great scenes of stellar evolution, indicated on sky maps and forming an unavoidable initiatory journey. All the acts of this play take place at the same time, right before our eyes, in the apparent quietude of the sky. Gestation, where stars are born in clouds of gas and dust . . . stellar families, clusters consisting of hundreds of stars born in the same litter . . . stellar life, spent alone or in couples and punctuated by unstable bursts of anger and eruptions . . . the stinginess of small stars and larger ones' overconsumption of energy . . . the planetary nebulae testify to the age and agony of stars, which die in the explosions of supernovae and enrich the universe with the debris of their infernal forges.

Moving away from the Milky Way, the sky becomes much blacker and less rich in stars, while the horizon becomes clear, letting sight extend outward into the profound abyss. Seen through a telescope, galaxies form cities with nebulous contours that come together in groups as if they were resisting isolation. How can you not imagine that, in the light emitted millions of years ago by these minuscule universe-islands, billions of systems similar to that of the sun are hidden? Tiny glimmers which provoke cosmic vertigo . . .

1.

THE STARS

Through a little or giant telescope, the stars all look like points. Due to the distances at which these suns are found, their disks cannot be magnified. All instruments do is increase their luminosity. Similarly, they seem strictly immobile in the sky when, in reality, the stars are moving. However, to catch even their quickest movements, you need to compare their positions on photographs taken several years apart. They're like individuals in a big crowd but whose faces we can't make out. The most remarkable stars make themselves known by their luminosity and color, as well as by the environment in which they evolve. Schematically, the

Lagoon and Trifid nebulae observed in Namibia through an Astrophysics 100/600 telescope.

HOW FAR AWAY ARE THE STARS?

Almost all the stars that make up the famous constellations are located within a sphere with a radius of 1,200 light-years (ly), centered upon the earth. Knowing that the diameter of the Milky Way is 100,000 light-years, they're relatively close by. At the speed of light—approximately 186,000 miles per second—six hours are needed to reach Pluto and ten to leave the solar system. Invisible to the naked eye, the nearest star, Proxima Centauri—is 9,000 times further away. This little red dwarf, 10,000 times less brilliant than the sun and which belongs to the Alpha Centauri triple star system, is 24,855 billion miles away, or 13,000 times the distance of the earth to the sun. A voyage of 4.23 years made at the speed of light is necessary to reach this star in the Southern Hemisphere. A cosmic flea jump, an abyss for humanity.

Within a radius of sixteen light-years around the solar system, sixty-odd stars have been counted. Altair in the constellation Aquila, Procyon in Canis Minor and Sirius in Canis Major are the most well known to observers. The latter, the brightest star in the sky (twenty-three times more luminous than the sun) is 8.61 light-years from the earth. Thanks to observations by the European satellite HIPPARCOS, the distances, positions, movements and brightness of exactly 118,218 neighboring stars are known with a great deal of precision. Interstellar travelers, to your calculators: the North Star is 430 ly away; the giant of Taurus, Aldebaran, shines with all its fire at 65 ly, Antares of Scorpio, "the devil's eye" at 700 ly. Vega of Lyra is a neighboring nightlight at 25.3 ly, while Deneb of Cygnus, at 3,000 ly, earns the title of "most distant star visible to the naked eye." If the Milky Way were reduced to the dimensions of a pancake the size of Paris, Proxima Centauri would be located only four inches away from us, and Deneb thirty feet! Beyond 5,000 ly, astronomers have trouble observing the stars of our galaxy: large clouds of gas and dust form a screen and filter out their light.

young ones are blue and bright, the old ones red and faint. Between the two, white and yellow dominate, like the sun. But if certain ones shine brightly, while others are at the limits of visibility, it's more often because of their distance than their maturity. And it's thanks to only several hundred stars, our closest and most brilliant neighbors, that the celestial diversity can be appreciated.

THE STELLAR WAY OF LIFE

For an initial intrusion into the life of the stars, the observation of a stellar couple makes sense. Contrary to our luminous bachelor—the sun—the majority of stars live in communities, and triple or quadruple systems aren't rare. Those which spend their lives as two—double stars—revolve around their common center of gravity and are so close together that to the naked eye they appear like one and the same star. The case of Alcor and Mizar—in the tail of the constellation of the Big Dipper—is original, for these stars rely on illusion to hide themselves better. Apparently, they seem to live side by side, and observers use them to test their visual sharpness. In reality, they owe their apparent twinness to a pure chance of position: they are very far from one another and form an optical couple brought together by an effect of perspective. But this illusion hides an actual binary system: that of Mizar which, magnified sixty times in a telescope, reveals a companion, the first double star identified back in 1650. Like rabbits pulled out of a magician's hat, astronomers discovered a third star in this system, a body called spectroscopic because, invisible, it lives hidden in the light of its two companions.

The most beautiful double stars are those of different colors. The Albireo system is a perfect illustration of this. These two little stars, in the head of the constellation Cygnus, are two of the jewels of the summer sky. The brighter star casts flashes of topaz blue into the night, while its companion shines with a curious yellow-orange light, between amber and sapphire.

A Star Often Varies . . .

The most attentive observers will notice as well that the luster of stars isn't constant. The light emitted by more than thirty thousand stars varies, increasing or diminishing over periods ranging from several minutes to several years. Certain stars pulsate, they beat like a heart and inflate and deflate with an exceptional regularity: they're the Cepheid stars. There exists a mathematical relationship between the pulsation period and the luminosity of these giants, which astronomers use to measure the distances of galaxies. Other stars can start erupting and increase their shine by a factor of one hundred thousand. These novae, from the Latin "new star," are stars reaching the end of life (the most massive explode into supernovae). There also exist false variable stars, which change periodically in luminosity for reasons of eclipses, provoked by the passage in the tight binary systems of a star over its companion's disk. It's in the surveillance of variable stars that amateurs contribute most to the collection of information destined for professional astronomers. The tracking of nocturnal beacons brings together thousands of observers throughout the world, and the "variablists" are, themselves, their own little family.

THE COMMUNITIES OF STARS

Clusters are another curiosity of observation. Here, too, you mustn't trust appearances. In certain regions of the sky—this is the case for the Milky Way—the stars are so numerous and close to each other that they seem brought together in groups and form, by effect of perspective, little packets. These stellar clouds or asterisms aren't composed of stars bound to one another like you find in open and globular clusters.

Open cluster of the Pleiades observed with the aid of a Takahashi CN 212 telescope, with 3.9 f/D.
[© *Ciel & Espace*/P. Durville]

Open Clusters

Open clusters are the product of a nebula (a cloud of gas) which has condensed into multiple lumps within which young stars have recently formed. Their light is vivid, blue, and illuminates the gaseous swaddling bands they haven't completely cast off. Five of the seven stars of the Big Dipper belong to the nearest open cluster to the earth. The Hyades, in the constellation Taurus, and the Pleiades are magnificent examples. In this latter cluster, whose arrangement evokes a miniature Little Dipper, six or seven stars stand out to the naked eye, compared to hundreds in a telescope. The faint gleam of gas and dust clouds that surround the most brilliant among them—in particular Merope—are visible only under exceptional atmospheric conditions.

Globular Clusters

These are primitive objects, contemporaries of the formation of our galaxy around 12 billion years ago. Made up of hundreds of thousands to several million old stars, they're distributed around the center of our galaxy and are the most extraordinary formations you can observe. In a telescope, their distance is so great (ten thousand light-years on average) that it's impossible to distinguish them from the stars. The Hercules cluster, for example, offers the grandiose spectacle of a compact and granular ball of stars, surrounded by peripheral suns. A voyage through the sky of the Southern Hemisphere is an occasion to discover the two most majestic of the globular clusters: Omega Centauri and 47 Toucanae. Astronomers estimate that the stars found at the center are fifty times closer to one another than the sun is to its most immediate neighbor, Proxima Centauri (see sidebar, p. 220). Is the sky still black there?

In the constellation Scorpio, right next to the brilliant star Antares, the globular cluster Messier 4 concentrates several hundred stars into one hazy packet.
[© *Ciel & Espace*/S. Guisard]

2.
THE NEBULAE

When the night is very black and pure, and only to eyes perfectly accustomed to obscurity, it's easy to locate with binoculars the little hazy and cottony clouds that float amidst the stars. Certain are faintly luminous—as if lit from inside—while others, on the contrary, mask with their darkness the stars behind them. Some of these nebulosities occupy very large portions of the sky; others, minuscule and very discreet, surround the stars like rings of smoke, Finally, ones resembling light veils or tattered spiderwebs floating in the wind seem to disperse into space in curls like smoke from a cigarette.

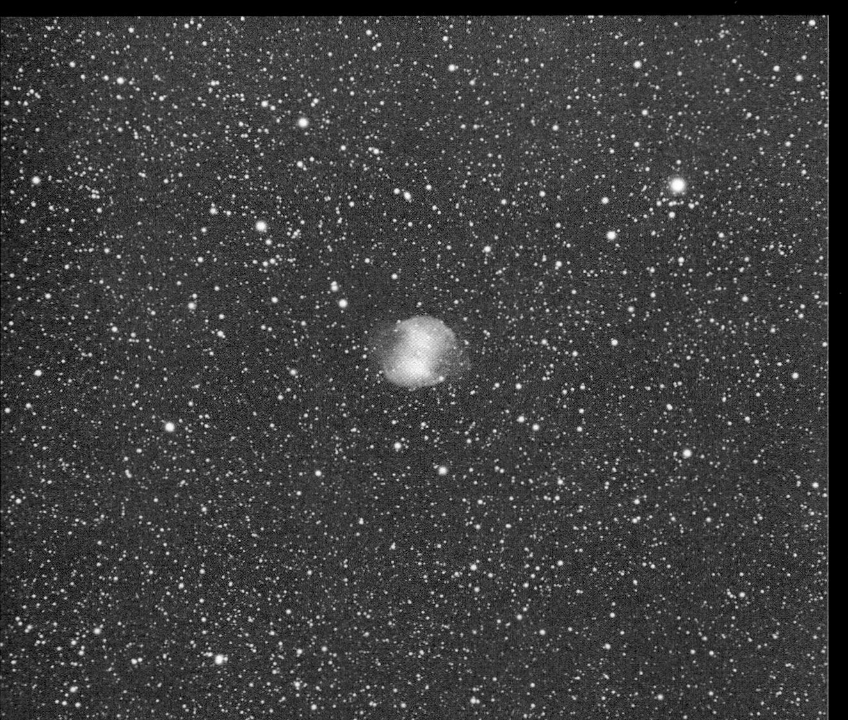

VARIETIES OF NEBULAE

Behind the word "nebula," which designates without distinction sets of gas or dust clouds, hides very different celestial objects. The nebulae called "diffuse"—like the large nebula of Orion or that of the Lagoon in Sagittarius—were recognized as being the breeding grounds of stars. These clouds emit their own light or reflect that of surrounding stars. The nebulae called "obscure" are veritable nests of dust that stand out like shadow puppets against the background of the sky. The planetary nebulae, called so because of their round and pallid appearance, are formed from gas ejected by dying stars which, in a final start, cast off their exterior envelopes. At the center of the most famous ones, like Lyra or the Dumbbell (also called Diabolo), it's theoretically possible to observe the star behind all the eruptions, provided you have a telescope of at least 200 mm in diameter handy. Finally, the remains of supernovae, like the veil of Cygnus or the Crab nebula, are the debris from the cataclysmic explosion of massive stars, clouds of gas and dust in rapid expansion.

HOW TO SPOT THEM

Observing a nebula isn't an easy exercise. Contrary to mind-boggling images transmitted by the operators of the Hubble Space Telescope, in the eyepiece of an

Planetary nebula of the Dumbbell, in the constellation of the Little Fox (Vulpecula).

[© *Ciel & Espace*/C. Ichkanian]

amateur telescope, the clouds of gas and dust stand out—due to their absence of contrast and colors! Worse, when you first locate them, they remain almost invisible. The secret to observing nebulae—and galaxies—resides in use of peripheral vision. Its principle is simple: during nocturnal vision, the cells of the eye most sensitive to light—the rods—aren't found at the center of the eye, but at the sides. Thus, to discover a faint object and make out its details, you have to shift your gaze laterally, look to the side and not move for a good five or six seconds, the time it takes for the light to accumulate. With a little practice, the habit of letting the detailed image of a nebulosity emerge in the periphery of your eye is fairly quickly acquired. Unfortunately, the cells sensitive to these weak luminosities don't know color: like a landscape weakly lit by the moon, the pale and diffuse objects appear gray. In the large, 400 mm telescopes, when light from the brightest gas clouds is strongly amplified, green or bluish flashes color the great celestial curtains. But the event is always exceptional.

Nebula of the Veils of Cygnus, gaseous tatters of the explosion of a star. [© Ciel & Espace/J. Riffle]

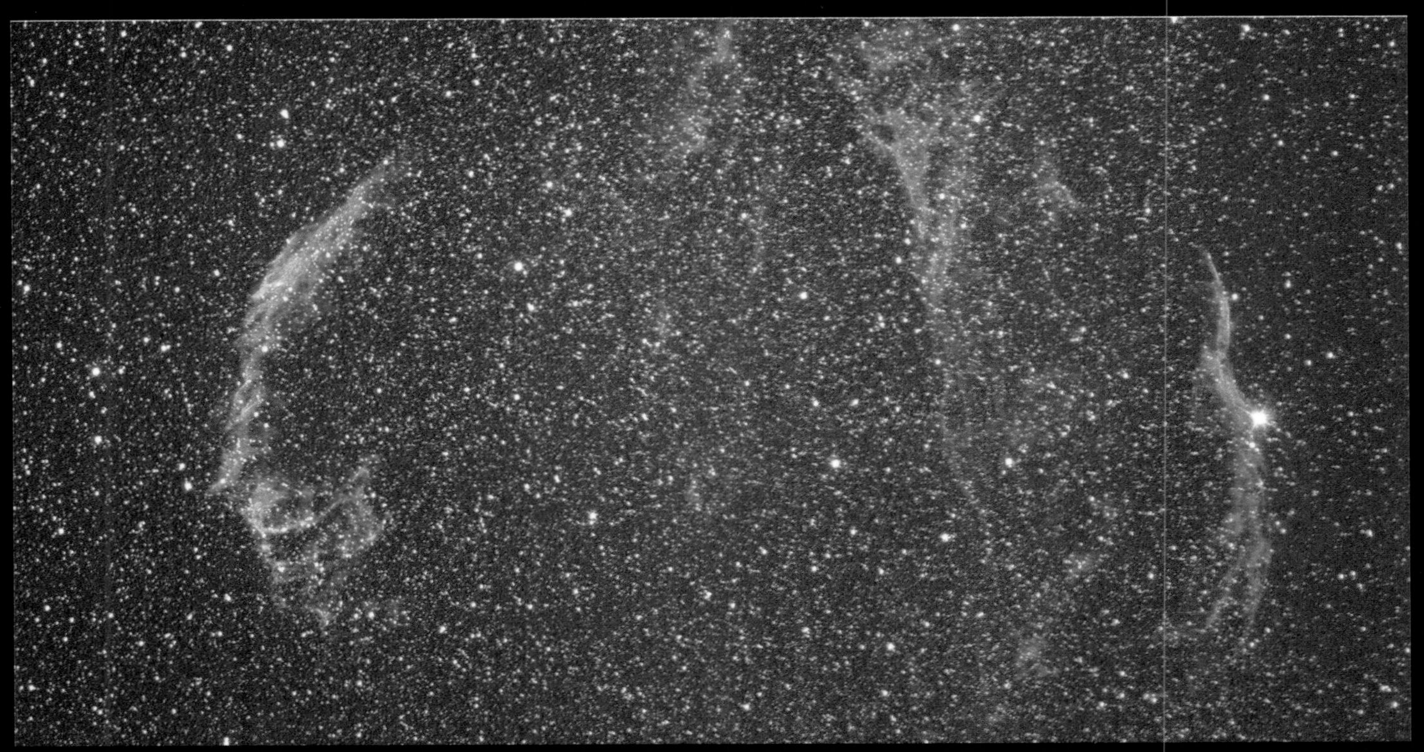

WHERE CAN YOU OBSERVE NEBULAE?

The list of beautiful nebulae is very long. Some have names—America, California, Rosette, Helix—but the majority are anonymous and have only a registration number, like a common car (see sidebar, p. 230). Throughout every constellation and in every season, it's easy to find these places of stellar evolution on sky maps and to attend, one eye pressed to the telescope, the birth and death of stars. Two different eyepieces are necessary to do so: the first, of weak magnification, offers a wide field of vision indispensable for locating the object; the second lets you magnify and bring details into focus, while causing the background of the sky to sink away.

In Winter

It's in the direction of the constellation Orion that your observation should be concentrated. The star of the show is the great nebula Messier 42, whose apparent diameter exceeds that of the full moon and which appears like a fuzzy spot to the naked eye. This cloud, whose shape is often compared to an albatross in flight, offers through binoculars a number of stunning details and reveals at its heart—through a telescope—

An immense cloud of gas and dust stretches across the entire constellation Orion.

[© *Ciel & Espace*/E. Beaudoin]

four bright stars in the form of a trapezoid. In reality, the nebula is only a small part of an immense gas and dust cloud that stretches across the entire constellation. In its folds, light from very young stars causes the gas and dust from which they emerged to shine, accentuating the system's depth.

In Summer

It's in descending the Milky Way—from north to south—that the magic of the spectacle occurs. From the constellation Perseus to those of Scorpio and Sagittarius, sky maps suggest visiting a large number of nebulosities. The shells of gas expulsed by dying stars, the fields of dust and the cottony breeding grounds of future suns will become familiar points of reference to a reader whose book—the sky—spreads all its pages out in no apparent order.

It's at the zenith, in the summer triangle composed of Cygnus, Aquila and Lyra, that the traces of enormous stellar catastrophes are most numerous. Planetary nebulae, remains of supernovas and bubbles of gas blown by shockwaves, are legion there. More to the south, where the Milky Way becomes dense with stars, beautiful, diffuse nebulae like those of Scorpio and Sagittarius (in the direction of the galactic center) remind you that the stellar machinery is always working and permanently manufacturing new stars.

Patience and subtlety are the keys to successful observation. In the eyepiece of a telescope where, apparently, you see nothing other than a black field speckled with stars, foggy clouds may suddenly appear in your peripheral vision. At first only a fleeting impression, then a certainty, the cottony traces of the birth or death of a star strike both eye and imagination.

The galaxy Messier 51, in the constellation of the Hunting Dogs, is a spiral of billions of stars, bound to an accompanying dwarf galaxy.
[© *Ciel & Espace*/C. Ichkanian]

3.
THE GALAXIES

From nebulae to galaxies, there's apparently only one step. Moreover, until the mid-1920s, astronomers labeled "spiral nebulae" these hazy little disks stripped of stars where dust clouds seemed to whirl. The proof, by Edwin Hubble, of the presence of stars in the Messier 31 nebula of Andromeda smashed to pieces our vision of the universe. To convince yourself, you need only locate, starting from the star Beta Andromeda—between Cassiopeia and this constellation—the pale and oblong spot representing the famous galaxy to appreciate the enormity of the discovery. At around 2.5 million light-years, a disk composed of at least 200 billion stars and innumerable gas and dust clouds

MAPS AND ATLASES OF THE DEEP SKY

To navigate the deep sky, you must choose a map—or atlas—adapted to the diameter of the instrument used and the faintest stars it lets you see. The naked eye doesn't discern stars exceeding magnitude 6. A small telescope of 60 mm lets you reach to magnitude 11 and a 115/900 telescope magnitude 12.4 (see sidebar, p. 170). A street map isn't designed for the same use as a state or national highway map.

In the same way, the usual maps of the constellations and their principal curiosities don't offer the same degree of precision as the *Millennium Star Atlas*—published by the European Space Agency from observations of the satellite HIPPARCOS—whose two volumes list the positions, distances, proper movements and real and apparent brightness of more than a million stars.

For amateurs, the most popular sky maps for reference are those of the American Wil Tirion. These are the *Cambridge Star Atlas,* designed for naked-eye observers and binoculars; the *Sky Atlas 2000.0,* which has twenty-six large maps itemizing 43,000 stars and 2,500 objects in deep space, and the *Uranometria,* a two-volumne work with more than 330,000 stars and 10,000 celestial objects. Among the latter, the most well known are designated by the numbers assigned to them in classic star catalogues. Those of Frenchman Charles Messier (1784), which counts 107 references, and of the Dane Johan Dreyer, who itemized a list of more than 15,000 diffuse objects in the *New General Catalogue* (NGC), are the most commonly used. For example, the galaxy Andromeda is registered on the maps as being at the same time "Messier 13 and "NGC 224." Finally, photo atlases which sweep the sky zone by zone can be the specialized complements to the star seeker's array.

Astronomy software, widely available in stores, has revolutionized the sector of celestial cartography. Today, amateurs can have at their disposal on computers stellar databases established for the Hubble Space Telescope. The "zoom" function lets you frame details of fields you've already explored with your instrument, and the number of visible stars is comparable to what you can see through the eyepiece of a telescope. And there's much less turbulence!

The large Andromeda galaxy (Messier 31), perfectly visible to the naked eye, observed through a Takahashi E-160 telescope.
[© *Ciel & Espace*/J. Riffle]

revolves around itself like spinning fireworks. This "universe-island" is the spitting image of our own cosmic condition, a twin sister of the Milky Way seen from afar. From very far, in fact: the light rays that reach us from it today left it back when man was still only a primate . . .

Beautiful Andromeda

Andromeda is, with the galaxy of Triangulum (Messier 33, see sidebar, p. 230), one of the rare superstars of the deep universe visible to the naked eye on a clear night. With a telescope, its central core and the periphery of its arms stand out sharply against the background of the sky, a bit like a luminous halo that betrays at night the presence of a distant, big city. As a general rule, seeking galaxies requires a good use of peripheral vision, a very luminous instrument—equipped with a low-magnification eyepiece—excellent sky transparency and low atmospheric turbulence. The magnitude, apparent diameter and angle at which these diffuse objects present themselves will help you locate them. When a galaxy is seen in profile and at a cross-section, its light is more intense than when it presents itself from the front. Similarly, its shape—spiraled, elliptical, lenticular or irregular—its size—from dwarf to super giant—and most obviously, its distance influence its conditions of visibility.

Where Can You Find Galaxies?

Taking off in search of galaxies requires shifting your gaze away from the plane of the Milky Way. Its density of stars and gas and dust clouds mask entire stretches of sky. Our own galaxy's disk prevents us from seeing worlds beyond it. As a general rule, you'll find them in three basic locations: near the Big Dipper, visible all year long and not far form the north galactic pole (the direction of the Milky Way's axis of rotation), near the constellations Leo, Virgo and Berenice's Hair, all visible in the sky in spring, and finally, in autumn, near Triangulum, Pegasus and Andromeda. In all, hundreds of galaxies are accessible to 60 mm to 200 mm instruments, and tens of thousands to larger-diameter ones.

MESSIER'S OBJECTS

The Crab nebula—the remains of a star whose explosion was observed by the Chinese in the year 1054—is Messier 1 (or M 1). The nebula Dumbbell, visible through binoculars between the constellations Arrow and Cygnus is Messier 27 (M 27) and resembles a little packet of hydrophilic cotton floating in space. The large cluster of Hercules is M 13, the nebula of Lyra, M 57, and the list goes on and on. One hundred and nine celestial objects, clusters, nebulae and galaxies form the upper crust of the most famous "Who's Who" in the world, "Messier's Catalogue." It was created by the seventeenth-century French astronomer Charles Messier. Nicknamed by Louis XV "the ferret of comets"—he discovered sixteen of them and observed more than forty—Messier took it upon himself to keep track of those nebulosities susceptible of being confused with comets. The first *Catalogue of Nebulae and Star Clusters,* published in 1771, described forty-five objects. It was increased a few years later thanks to Pierre Méchain (one of the surveyors of the meridian), who added sixty-fold references. Its ease of use—objects are designed by the letter M, followed by a number according to their order in the catalogue—as well as the mnemonic tricks it provides for learning its numbers by heart and entering a circle of initiates—has ensured its success up to the twenty-first century.

Each year, around the spring equinox, the greatest observers in the Northern Hemisphere organize a contest of visual dexterity, baptized the Messier Marathon. It consists of observing from sunset to sunrise, on a moonless night, the greatest number of objects possible referenced in the famous astronomer's catalogue. In principle, one night should suffice, provided you don't come face to face . . . with a comet.

Messier 1, the Crab nebula, is the residue from the explosion of a star. [© *Ciel & Espace*/P. Durville]

The Galactic Continents

Galaxies, too, hate solitude and group together in clusters counting hundreds, even thousands of members. The Milky Way is no exception and, in the company of the galaxies Andromeda, Triangulum, the Magellanic Clouds (visible in the Southern Hemisphere), as well as thirty-some other partners, forms a system named the Local Group by astronomers. Within this set, the galaxies are all bound to each other gravitationally and rotate in clusters. But it's much further away, hundreds of millions of light-years from here, where the architecture of these vast sets is revealed. In a region relatively empty of stars, between the constellations Leo and Virgo, a group of galaxies, known by astronomers of the seventeenth century as the "nest of nebulae," cracked open the door of the cosmological universe, of space seen on the large scale. The Virgo cluster, a vast set composed of several subgroups, is around 50 million light-years away. Twenty-some galaxies are accessible to a 60 mm telescope. Out of the three thousand objects hidden in this region, nearly a hundred of them are at the grasp of a 200 mm telescope. Present in the shape of cottony masses, these nests of other worlds reveal no detail to even the sharpest eye. Worse! Depending on eye fatigue and turbulences, they are appear and disappear in your peripheral vision and, like phantom images, sometimes make you doubt their existence. "It's a very fertile field," wrote Camille Flammarion in *Les Étoiles*, "the space which stretches across the sky from the Ear of Virgo in the south up to Berenice's Hair in the north. [. . .] It's the richest region in nebulae: there are more than five hundred of them. Seeds of future worlds, germs of the universe to come, fertile jelly which seems to tremble, suspended on the ethereal ocean! The number of nebulae eclipses that of the stars!"

The famous popularizer had no idea how well he'd put it. In the course of one of the longest photo shoots ever taken of the sky—130 hours straight, a bit more than five days—the Hubble Space Telescope photographed within a celestial field as tiny as a pinhead galaxies 10 billion times less luminous than the faintest stars visible to the naked eye.

Of the two thousand known galaxies, among the furthest away in the universe, astronomers deduced that the entire galactic population could reach as high as 100 billion galaxies. Much more than all the stars visible individually through a telescope!

5

Astrophotography

1. FIRST STEPS 2. INSTRUMENTS AND PHOTOGRAPHY 3. DIGITAL ASTROPHOTOGRAPHY

The very first astronomical image was a daguerreotype: a first quarter moon made its impression on a sheet of silver-plated copper. The surface of our natural satellite, with its mountains, seas and craters, is perfectly visible. In that year, 1840, it looked no different than it did at the dawn of the third millennium . . . except for a set of boot prints! This image, made by the American John William Draper, was a veritable feat. It wouldn't be exceeded until 30 September 1880, when his own son, Henry, succeeded in fixing Orion's large trapezoid onto a sheet of wet collodion. After fifty minutes of exposure, one of the heroes of the deep sky offered the brand new field of astrophotography its pedigree. The project of the photographic map of the sky, launched in Paris during the International Astronomical Congress in 1887, established it as the new method for studying celestial objects. And, in 1929, snapshots of the sky changed the face of the world by allowing American astronomer Edwin Hubble to discover the expansion of the universe.

1.

FIRST STEPS

A note to the astrophotographer: don't tread lightly. Photographing the sky is a veritable art form, and its practice requires a level of obstinacy, patience, and calm that isn't within in everyone's grasp. You have to master its tools—both camera and telescope—learn how to work in absolute stillness and blackness, be willing to grope around in the dark, and fail over and over again before finding the right exposure times, film and methods for focusing. There isn't any secret to creating a good image; astrophotography is learned on the job, and a beautiful photo of a planet or galaxy is much more a question of training and experience than technique.

What Materials Are Needed?
The neophyte hoping to taste the thrill of the hunt for the night beauties has every reason to approach matters progressively. For an initial approach to astronomical photography, a sturdy photo tripod, a 24 x 36 or 6 x 6 reflex camera offering B or T exposure (this lets you control the exposure), a lens and a cable release will do the trick. Once the film is loaded in the casing, the diaphragm open to its maximum, the focusing trained on the infinite, the photo shoot can begin. No tricks exist to assure capturing the superstars of the spangled vault on film. Obstacles, however, are legion: stray light, turbulence—atmospheric and instrumental—wind, vibrations of the tripod or reflex mirror, mist. A multitude of inherent barriers to capturing the world of the night play spoilsport. To tilt the odds in your favor, you need to choose your site carefully (just like in observation). You also need to make sure all the requisite atmospheric conditions are met—very black sky, limpid atmosphere, etc. (see chapter 1).

The Stars' Rounds
The most spectacular image for a beginner is without dispute that of the displacement of the stars around the celestial North Pole. The recipe is simple: mount your photo casing upon a good tripod and attach a very open, wide-angle lens (28 mm or 35 mm at f/2 or 2.8). Direct the lens toward the sky and establish in the bottom of your frame a landscape to serve as a foreground—a range of trees, the silhouette of a village, a clock tower, etc. Next, center this same shot upon the faint glow of the North Star and begin the exposure. Carefully emphasize the foreground using several handheld flashes and wait a good hour, so the earth has time to turn a bit. Once the film is developed, the photographer will discover what the eye is completely incapable of recording: a circumpolar star. This is a series of luminous trails in the form of arcs, which encircle the celestial pivot of the North Star like a crown. With a little luck, the flash of a shooting star will break this beautiful harmony with an ill-tempered streak, like speed thumbing its nose at inexorable slowness.

2. INSTRUMENTS AND PHOTOGRAPHY

To use a telescope to explore new photographic territory is very tempting, though especially arduous: a photo of heavenly bodies is nothing like a wedding photograph, even a successful one.

The amateur who insists on capturing marvels on film as seen by the eye through a telescope has two solutions. The simpler one involves taking the photo as described above, but only after placing the camera on the tube of the telescope to benefit from the mount drive. The device moves along with the sky, allowing you to remain fixed on a star and avoiding the appearance of ghost images. This isn't easy: if the mount shakes a bit or has the slightest failure, the desired object will end up looking like a comma.

CCD camera fixed to the focus of a Schmidt-Cassegrain telescope. [© SPJP-Paralux]

The second possibility involves replacing the lens of the camera with the telescope, which will then serve as a super-telephoto lens. The casing, without its lens, is attached to the back (to the side for Newton-type telescopes) of the instrument with the help of a specially designed ring. Photographers can choose to remove the eyepiece or to leave it on if they want to enlarge the images. To focus using the focusing thumbwheel, you have to look through the camera's viewfinder. The exposure is activated when the sky is stable and the image sharp and centered. Do your best to avoid shaking the instrument.

Exposure Times

Exposure times are not at all the same depending on the desired object: from the order of several seconds for planets, to ten to forty-five minutes for stars depending on the instrument, the sensitivity of the film, the state of the sky, and so forth. (see sidebars, this page and p. 236).

Here, too, it's useless to hope for a miracle trick. Astrophotographers have all learned by trial and error. Starting with an average value and working around it, taking a series of snapshots of the same object using different exposure times (don't forget to note them carefully), you can better address the problem.

A Portrait of the Moon

Because of its strong luminosity and size, the moon is the ideal subject for a first night combining photography and telescopes. An image's dimensions and sharpness of resolution depends on the telescope's focal length. If the moon is photographed with a 50 mm lens, its diameter on film will appear only 5 mm across. With an optical device of

FILM

Whether black-and-white or color, good film for astrophotography is abundant. It's hard to recommend one type over another, as users all appreciate different types of rendering, contrast and color. Two qualities, however, are essential: a sharp grain for image contrast, and a maximum of film sensitivity for reducing exposure times. The only problem is that the twain never seem to meet.

It seems evident that, for luminous celestial objects like the moon and planets, the use of slow, lower-sensitivity film (50, 100 ASA) is necessary. The more you magnify the image, the more the exposure time will increase and the faster the film will have to be.

Sensitive film such as TP 2415 by Kodak, T-Max 400 to 3200 or film ranging from 800 to 1600 ISO is designed for photographing nebulae, galaxies or gas and dust clouds of the Milky Way. Here, too, the matter isn't simple: after several minutes of exposure, sensitivity falls drastically. This effect, called reciprocity failure, is corrected with film for scientific use (type 103 aE, aF and aO) but such film is fairly grainy. The best solution consists of testing several types to compare their respective merits, adjusting the exposure times according to the equipment used and the object photographed. The rest depends only on your taste.

ıc meter focal length, the moon's image will grow to ›out a centimeter. The conclusion is obvious: to dive to the heart of the lunar surface and capture the details ' its seas and craters, you need long focal lengths and gh magnification at your disposal. What you see rough the lens is what you'll end up photographing, › it's up to you to choose an eyepiece appropriate to ›ur instrument and desired magnification.

The entire secret behind capturing a good image ' the moon is in the focusing, which must be perfect, › playing with the telescope's adjustment knob. hen the image is magnified, it's at the edge of the oon or its terminator—the line of separation between e lit part and the obscure part—where the adjust- ents can be most easily refined. As a general rule, rbulence plays spoilsport and makes this operation ›ry delicate. All you can do is wait and keep your eye ıt for a hole in turbulence, a moment of calm lasting veral seconds during which the image can be stabi- .ed. Rare and precious instants, which you exploit to e maximum by eliminating all sources of instrumen- l vibration. When the photographer presses the shut- r release, for example, the viewfinder mirror lifts up ıd can cause the instrument to shudder. The shot will : blurry. Ideally, you should develop your own film so u can have access to all your photos, including the urry and overexposed ones. This will make you more vare of your errors and help avoid them in the future.

otograph of the focus of an astronomical telescope.
[© SPIP-P2r2lux]

3.

DIGITAL ASTROPHOTOGRAPHY

Until recently, astrophotographers have relied on the same techniques as the Sunday reporter did, or classic silver-salt photography. They only needed a camera, a lens and some film. The latter has served since the invention of photography as the physical support of the image, whether it be in the sky or on the earth. Now, a new technique which, according to its partisans announces nothing less than the death of film, is in the process of revolutionizing the little world of astropho- tography: the Charged Coupled Device (CCD) Camera.

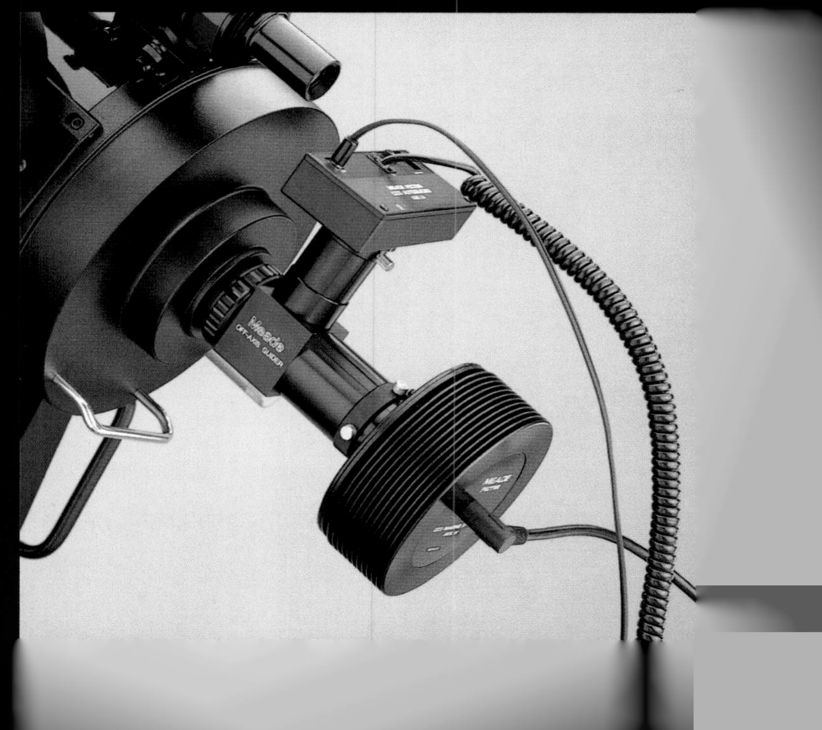

EXPOSURE TIMES

For stellar photography, exposure times don't depend on the diameter of the instrument but its aperture ratio f/D (focal length divided by the diameter). The weaker this is, the more the duration of the shoot will be reduced. If it's inferior to 10, for example, the exposure will take on the order of thirty minutes to an hour.

In the case of the moon and planets, the exposure is a function of the transparency of the sky and the body's luminosity. From 0.5 to 2 seconds for the moon based on the time of lunation and the region photographed, 0.5 seconds for Mars, and 2 to 3 seconds for Jupiter are base figures around which the astrophotographer can "fudge," that is, test superior and inferior values.

HOW MUCH DOES A CCD COST?

There exist three families of CCD cameras. The first, from $500 to $1,200, lets you become familiar with digital telescope photography and can be used with your instrument's automatic guidance control. The second, from $1,500 to $3,000, offers powerful and sensitive imaging capability. Finally, the third is reserved for well-to-do lovers of professional-quality cameras.

A choice of exceptional cameras range from $5,000 to more than $23,000 and offer, at the top of the scale, fields comparable to those of 24 x 36 film. That gives just a hint of revolutions to come. Far out!

developed in 1970 by Bell Telephone Laboratories i
the United States.

Detectors of this type used by amateu
astronomers function according to the same princip
as webcams (small Internet cameras) and commerci
digital cameras: the film is replaced by a little pellet,
photosensitive surface that contains a pile of little si
icon chips, or pixels. The CCD is inside a sort of bo
stripped of lens and sight system, which star hunte
affix to their instruments like normal cameras. Th
chips capture the light emitted by the stars and co
vert them into electrical impulses. The latter are the
transformed into digital data, which are transferred t
the computer linked to the CCD camera. This is wh
drives the focusing, and the image appears on scree

All-Purpose Machine

Captured by the CCD and saved in the memory of th
machine, this image can then be processed by softwa
with very large use capacities: you can eliminate defect
combine the shoots for "composing" a fine and co
trasted image, simulate a long exposure equal to the su
of multiple exposures, put together mosaics to represe
the objects in a wide field or animate the movement
clouds across the atmosphere of Jupiter—the palette
possible manipulation never stops getting bigge
Asteroids, comets and supernovae are much easier t
spot using software developed with the CCD astronom
the computer compares images of the same patch of s
taken at different moments and detects the stars th
have moved or are new. The computer can also b
attached to the mount of the telescope and control i
movements: it's now the computer that's in charge
making the instrument move and keeping the target sti

Large nebula of gas and dust from Orion.
[© *Ciel & Espace*]

The Best of Both Worlds?

On a practical level, the advantages of the CCD are immense. Contrary to film (see sidebar, p. 234), its sensitivity is constant. It captures certain infrared light, reduces exposure times, the photo is available immediately, offers unlimited attempts and replaces printing with free clicks of the mouse. If the quality of the observer's telescope, mount, sky and know-how remain high, the CCD lets you correct, over the course of the night, the most obvious image errors.

CCD cameras do have their disadvantages. They're complex and expensive. For example, for the detector to function properly, it has to be cooled (to a temperature from -40°F for cameras available on the market, to -184°F for specialized ones) using a special device. Users complain of difficulty focusing, of having to take test images for calibration, and of the faintness of the visual field—so many arguments against a costly technology that links observer to computer on the field.

But the frontiers are blending together; the webcams and the digital casings that weren't developed for astronomical photography are being turned toward this goal by their users, whose targets of preference are the moon, planets, and other shiny light sources. Those who wish to stick with silver-salt, black-and-white and color photography can also digitalize their wide-field snapshots and compose on their computer, with the help of ad hoc software, new and splendid images of the deep sky. All are available on the Internet, the river of the digital universes.

Acknowledgments

We wish to thank Mr. Jean-Eudes Arlot, director of the Institute of Celestial Mechanics, and Mr. Patrick Pelletier, optical engineer, for their attentive reading and sensible advice.

Originally published in France by Éditions du Seuil / Association française d'Astronomie in 2001 under the title *Clés de voûte, savoir l'astronomie, voir le ciel.*

Original ISBN 2-02-041533-X

Library of Congress Cataloging-in-Publication Data available.

ISBN 2-02-059692-X

English translation by Peter F. DeDomenico
English typesetting by Janis Reed

Cover design by Benjamin Shaykin
Cover photograph by Paul and Lindamarie Ambrose/Getty Images

Manufactured in Italy

Distributed in Canada by Raincoast Books
9050 Shaughnessy Street, Vancouver, British Columbia V6P 6E5

10 9 8 7 6 5 4 3 2 1

Chronicle Books LLC
85 Second Street, San Francisco, California 94105

www.chroniclebooks.com